T0253797

The Mystery of the Seven Spheres

Giovanni F. Bignami

The Mystery of the Seven Spheres

How *Homo sapiens* will Conquer Space

 Springer

Giovanni F. Bignami
Italian Institute for Astrophysics
INAF
Rome
Italy

Translation by Alessandra Scaffidi Abbate

ISBN 978-3-319-17003-9   ISBN 978-3-319-17004-6   (eBook)
DOI 10.1007/978-3-319-17004-6

Library of Congress Control Number: 2015936153

Springer Cham Heidelberg New York Dordrecht London
© Springer International Publishing Switzerland 2015

Translation from Italian language edition: *Il mistero delle sette sfere* by Mondadori, © 2013. All rights reserved

This work is subject to copyright. All rights are reserved by the Publisher, whether the whole or part of the material is concerned, specifically the rights of translation, reprinting, reuse of illustrations, recitation, broadcasting, reproduction on microfilms or in any other physical way, and transmission or information storage and retrieval, electronic adaptation, computer software, or by similar or dissimilar methodology now known or hereafter developed.
The use of general descriptive names, registered names, trademarks, service marks, etc. in this publication does not imply, even in the absence of a specific statement, that such names are exempt from the relevant protective laws and regulations and therefore free for general use.
The publisher, the authors and the editors are safe to assume that the advice and information in this book are believed to be true and accurate at the date of publication. Neither the publisher nor the authors or the editors give a warranty, express or implied, with respect to the material contained herein or for any errors or omissions that may have been made.

Printed on acid-free paper

Springer International Publishing AG Switzerland is part of Springer Science+Business Media (www.springer.com)

Contents

Chapter 1
Introduction

Like many others, while at high school I was very keen on Luigi Pirandello (and in love with a classmate as well). One of his short stories, *The Tragedy of a Character*, struck me in particular. In it, Pirandello tells us how, on Sunday mornings, he used to create his ideas for characters to be used in his future novels.

Inspired, I decided to follow his example. I would meet "my authors", i.e. those writers who, from time to time, I liked rather more than others. On a Sunday morning, and maybe at other times too to tell the truth, I would go to a shelf and pull out a book of one of my favourite authors, look at it, smell it, thumb through it and hold it in my hand. I would do this for love, for inspiration or just for the joy of it. With a book in hand, it was natural for me to talk to the author; in particular, to ask him how he came to write such a beautiful thing. Few authors answered me, or maybe I did not understand, and almost always I would put the book back on its shelf without thinking any more about it.

Even now, half a century later, I still occasionally have such encounters (in a very physical way) with the books of my favourite authors. By now, there are many more books on my shelves, and they have undergone much change over the course of time. The other morning, a Sunday, and (who knows?) perhaps because it was a Sunday, I felt I wanted to meet again with Jules Verne, my absolute favourite author as a youngster. I own an old illustrated edition of *A Journey to the Centre of the Earth*, an impossible story that had always fascinated me. It is a book with an unmistakable and unforgettable scent of its own. I opened the book by chance at the page describing the Runic manuscript that will guide the intrepid Professor Lidenbrock towards the centre of the Earth, and again I was overwhelmed by the story.

A few words concerning Jules Verne, for the benefit of those who are less fanatical than I am about the famous French writer. He was a Frenchman, born in 1828 in Nantes, the port of sardines travellers; at the age of eleven, he runs away from home to embark on a ship to the Indies, but his father, a leading man of the law, finds him and takes him home. Forced to finish school, he becomes a lawyer, although reluctantly. He then unleashes in writing his passion for exploration, despite the legal career which he had had forced upon him by his father. For years,

© Springer International Publishing Switzerland 2015
G.F. Bignami, *The Mystery of the Seven Spheres*,
DOI 10.1007/978-3-319-17004-6_1

he divides his time between work and writing plays; up to the age of thirty five, he had only written *Five Weeks in a Balloon* (although throughout his whole life he only spent a grand total of 24 min on a balloon) and then, in the space of two years, he wrote, *Journey to the Centre of the Earth* and *From the Earth to the Moon*, and immediately found fame in 1852, the same year Napoleon III became Emperor. During the decade, from 1865 to 1875, he wrote four other masterpieces: *The Children of Captain Grant*, *Twenty Thousand leagues Under the Sea*, *Around the World in Eighty Days* and *The Mysterious Island*. Globally, Verne had a formidable literary success, which allowed him a life of great wealth (he bought, among other things, two yachts). As a great gourmet, he had an intense relationship with food, which is reflected in the extravagant menu prepared or tasted by his characters, from the very elaborate meals enjoyed by Captain Nemo's mariners to the simple but nourishing food of the castaways in *The Mysterious Island*. Other works followed, mostly novels, eighty in all, to describe approximately 64 voyages, at a time when one wrote with a quill pen by candlelight. More or less over the same period, from the 1870 (with the Germans in Paris...) to 1880, four ponderous volumes were published on the history of the exploration of the whole Earth, its continents and oceans. By this time, he had become so famous that in 1884 he was able to visit Pope Leo XIII, who seems to very much appreciate the French writer. In 1902, he meets the first great French film-maker, Georges Méliès, who immediately shoots *A trip to the Moon*, inspired, of course, by Verne. In 1904, a year before his death, diabetic, nearly blind and paralysed, he collaborated on the screenplay of another Méliès film and adapted from a play of 1882 by Jules Verne and Adolphe d'Ennery, *Journey Through the Impossible*: what better title for the last work of Verne?

Returning to my experience the other morning: after an hour of feverish reading, I was almost in a trance, when suddenly and without warning, Jules appeared at my side in front of the bookshelf. I seemed to be dreaming (and perhaps I was dreaming...), but it does not matter, the opportunity was too good. He was accompanied by his odoriferous cigar from which he was inseparable and he seemed well prepared and in the mood to talk. What does one ask one's living legend when he stands in front of you? What does a mathematician ask to a resurrected Gauss? Or a philosopher to Aristotle, perchance met in a bar? I threw myself in headlong.

"Jules" I asked, "May I call you Jules?... You made up all these incredible adventures—underground, under the sea, in the air, in space, even in time—you made them come alive to me and to millions of people all over the world. How did you do that? What is the secret of your narrative technique? How do you enchant and entrap the reader, as you have just done with me, yet again?"

"*Tu sais*", he replied, "it is not difficult. There are no special narrative techniques, no secret invocations of complicity or of suspension of disbelief with whoever might be holding one of my books. Simply, leave it to the reader, let them dream and then let them find written on the next page what they had dreamed of when reading the previous page. I mean, I've never written that stuff that you now call science fiction, my stuff was all perfectly believable, designed by engineers, and it was credible because it came of your desire to believe it. *Et voila!*"

"Your Secret" I said, "has the simplicity of genius, *mon cher*, *Jules*, and I hope you allow me to copy it a little, in a small... Actually, I had in mind to write something about the mania that man has, has always had, and always will have, for exploration. I was just looking in your books for a way to make the subject understandable, orderly, and attractive at the same time: I would like to mention man's constant path towards the exploration...

Here Verne interrupted me: "*Justement, mon cher ami*, You know that after 1870 I wrote a long treatise in four volumes on exploration... but having thought about it, truth to tell, I'm not quite satisfied with it...I think it came out a little... dry... and, well, seeing that I cannot change it, at least let me enter into this adventure with you".

I turned pale at the thought: "How so? Jules, you with me? This is too great an honour". With a good-natured smile, Verne said: "*Allons-y*, come along—let's try..." and with that I felt his hand on my shoulder.

A shiver ran all through my bones: Jules Verne had touched me! Reality or dream—it does not matter. Since that moment, I was never the same person: I was refreshed in spirit; every idea, every page seems to me to have been validated, if not suggested, by Jules, and everything is easier. An initial idea that we (!) had considered was first to talk about the birth of man's passion for exploration, but then, more importantly, to talk about what there is still left to explore and how we could do it today, tomorrow and the day after.

Jules, I must say, it is very well-versed concerning the extent of the exploration of the Earth's surface; think of the saga of *The Children of Captain Grant*, in which a group of adventurers go around the world in search of a disappeared hero. Verne, however, died a few years before both the North Pole (Peary 1909) and the South Pole (Amundsen 1911) were reached, and almost half a century before the last symbolically inaccessible place on the planet, the summit of Mount Everest was trod on in 1953. Moreover, Verne in 1870 knew what was then possible to know about underwater exploration. So when I told him what we know today about the centre of the Earth, the oceans or, even more, outer space, beyond the Moon—that was enough: a partnership was born: Jules would guide me through (he suggested it to me) the "conquest of the seven spheres" I would simply explain what we know today and what we hope to be able to do tomorrow. Jules, however, promised to supervise all the work and, most importantly, to send me by way of help, his imagination, where my own may not be enough.

I shall now try to explain what the seven spheres are, at least as I understand them. This one of the spheres is also one of Verne's ideas, passed to me with the generosity of a great gentleman. Even the term "conquest" is his: I probably would have preferred "exploration", but it is hard to argue with a great writer. But perhaps the difference is nuanced in time: if today one plants fewer and fewer flags, and tomorrow still fewer, maybe it is because what really matters is the exploration itself, rather than the achievement of goals, i.e. the "conquest".

The nearly spherical surface of the Earth with (its land to roam on and its seas to navigate) was for mankind their first field for exploration, and we shall call this Sphere 0, or the sphere of the land and sea. Today, there is not much mystery to

Sphere 0, but that is where we learned to explore (and to plant flags, often with disastrous consequences).

Going downwards, under miles of water, we come to Sphere −1, or the sphere of the deep sea and ocean floor. Today, we can do much better than Captain Nemo, the commander of the Nautilus, the "electric" submarine of *Twenty Thousand Leagues Under the Sea* which was so innovative for its time. Sphere −1 is essentially still being explored, and we will see how to do it, and why.

Further down, we find what we shall call Sphere −2, or the sphere of the underworld, the deep interior, the totally unexplored part of planet: here the famous *Journey to the Centre of the Earth* will be set in modern times. It will be an exploration of something that holds big surprises, we shall discover, and an exploration which could begin by using some incredibly innovative means.

It is inevitable that today we observe the Earth as entering a new geological era, which many have called the "Anthropocene". The term seems to us one that well defines a period with environmental conditions that have been altered by human activities, since modern *Homo sapiens*, with all their strengths and weaknesses, became the dominant ecological force on Earth. Thus, even if Spheres −1 and Sphere −2 still remain (relatively) spared by these onslaughts, imagining a future in which we will need to explore more and more seriously other planets to which we might need to travel is unavoidable.

Moving away from the surface of the Earth, we shall go into space, and here comes the good part. We immediately meet Sphere +1, or the sphere of Terrestrial Heaven: it is the strip a few 100 km above the Earth, where today thousands of satellites orbit and where hundreds of astronauts come and go all the time. Jules had overlooked this sphere a little bit, since in his time it was still too unknown. For us, however, Sphere +1 is increasingly used as a region in which to conduct fundamental science and test its applications, and it will become even more so, as well as a place in which to train ourselves to go farther into space.

Sphere +2, or the sphere of the Moon, is a thousand times farther from the Earth than is Sphere +1, and a thousand times more difficult to reach. Yet we did reach it half a century ago, and it remains a great historic undertaking. On the Moon and nearby, there is still much to do. I'm sorry, Jules, *From the Earth to the Moon* is a wonderful work of fiction... but to get to and from the Moon was not so simple, as you imagined as we shall see. However, chemical engines will still be all we need to travel on.

Sphere +3 is that of the "outer" planets of the solar system, i.e. that which starts with Mars, a thousand times more distant than the Moon, and continues as far as the Trans-Neptunian Objects. Here, we have to shift gears, as we will see, and concern ourselves with those forms of energy that will allow us to get there. Jules will be of little help here, since this is where nuclear physics will be of service and, indeed, at this point, we will leave him, probably going over modern books to understand nuclear propulsion.

He himself told me: "From now on, beyond the Moon, I can be of no more help, *hélas*. We need a professional: do you not know someone?"

Luckily yes. For Mars and beyond, we will be accompanied by a hero of real space exploration, Wernher von Braun, the (real) conqueror of the Moon; one who perhaps had less imagination than Verne, but who had more precise ideas on rocketry. We must be able to get hold of him in some way.

If we decide to use the right form of energy, and we already know that it is the nuclear one, we will see how the exploration of Mars is within our reach. And, inevitably, at that point, we will take off, just as the Phoenicians, the Vikings or Columbus did when they had at their disposal the right vessels. Not only that: but with a "little" extra effort, necessary, to achieve the right scientific and techno-logical know-how, we may extend our reach a 1000-fold and penetrate the thick-ness of Sphere +3. We could arrive at the edge of our Solar System. Brace yourselves, Jules and Wernher: here, we will even need to go a step further, and study thermonuclear fusion…

Just like the Aristotelian system had one final sphere (which Aristotle called of "the fixed stars", not knowing that they happily move around…), we also have a "seventh sphere", our Sphere +4, which we will also call "the fixed stars", in keeping with tradition. It is a sphere that has a radius of about twenty light years from us, and it is one that will take us into our galactic neighbourhood. It contains at least 131 stars around which we now know that numerous planets are orbiting. Indeed, many of them we have already discovered. They are very close to us and also very interesting: a good reason to go there as soon as we are able.

We do not yet know *how* to do it, but we already know that we would like to go there; we shall need one form of energy so efficient to make thermonuclear fusion look like petrol. Here, we have to work imaginatively and will do so in the company of and be inspired by someone who did not joke as to how much imagination he had: Titus Lucretius Caro, the Roman poet who, in the first century BC, wrote more than 7400 beautiful hexameters in six books entitled *De rerum natura*, where he describes more than The Nature of Things.

We will find the right method of propulsion strictly respecting the physics that we know today (because here we are not, unfortunately, in a Dan Brown novel…), just as Verne did when faced by problems bigger than he could imagine. However, even if we do not know how to be able to pierce the sphere of the fixed stars, we at least know that we have a strong motivation to do so: we will see that we have already discovered that some of the planets orbiting stars in our neighbourhood could be habitable (or, indeed, inhabited?).

Bon voyage everyone, and Jules, Wernher, Titus, please help me stay with me… And if seven spheres seem too many to our reader, consider this: the original idea of the celestial spheres was developed by Eudoxus of Cnidus, who, in the first half of the fifth century, guessed there to be 27 spheres, to account for the motion of the planets and the stars. Aristotle, half a century later, stole the idea, but with his mind more akin to that of an engineer (for a philosopher…), he wanted to imagine the spheres in more concrete, physical terms: as made of crystal, or something similar. Except that, by doing so, one confounds further a complicated mechanical system and Aristotle come to need as many as 55 spheres rotating and oscillating… We, after all, shall necessary in theory manage with far fewer.

Part I
Our Earth, Our Oceans, The Sky, The Moon

Prologue 1

A large bonfire roars in the Afar Depression, Ethiopia. Usually, the Depression is one of the hottest places in the world, but that night in November 1974, it was distinctly cool. Sitting around the fire, with a glass in hand and tears of joy in their eyes, there are a dozen otherwise very serious anthropologists—men and women from the most important universities of France, the USA and England. They are celebrating what may be one of the most important discoveries in the history of mankind: a fossil that will cast new light on the ancient origin of man. Donald Johanson, the curator of the Museum of Anthropology in Cleveland, is telling for the umpteenth time to Yves Coppens, of the Collège de France, how, while wandering in the dry bottom of a river in the afternoon heat, he saw a piece of bone protruding, and then another, and then another.... . Even Yves' eyes sparkle as he raises his glass of champagne brought from Paris by Land Rover just for occasions such as this (and never mind if it is not chilled enough ...).

At the end of the day, everyone realises that, incredibly, no duplicate bones have been found. In other words, all the collected bones belong to one and the same skeleton! Later, after months of analytical work, they will also realise that a record 40 % of the entire skeleton had been recovered. For the moment, they are celebrating around the campfire, while a tape recorder plays a great hit of the Beatles at the time: "Lucy in the sky with diamonds". They know that the skeleton is of a female and, inevitably, someone (nobody now remembers who) proposes to call her Lucy. All approve and toast again Lucy, a young female hominid, who clearly lived a long time ago.

Thus was born the name Lucy for the skeleton, officially known as *Australopithecus afarensis*, code name AL 288-1, although few people know this. They are the bones of what, at least symbolically, is considered to be the mother of us all (as well as a lot of other primates ...). We now even know Lucy's age. Not her age when she died, which by our standards was probably very young, when she was, perhaps, overwhelmed by the flooding of the river bed where she was found. But what is more important, we calculated the age in which she lived: a respectable 3.2 million years ago, with an error of less than 100,000 years.

We know several other things about Lucy: she was tiny, weighed 29 kg and was just over a metre tall, and she had a small head which was decidedly simian. But she also had one other very important feature: the shape of the bones of the pelvis and of the legs showed that she possessed the miraculous ability to walk upright, although with a little sway, so the experts say. Born and raised in the heart of Africa, she was, more or less, the first "quadrumane biped". She must have figured out (good for her) that standing up allowed her to see farther, above the tall grasses of the savannah. This greatly improves the chances of survival: for example, it allows you to see the lion a few seconds before it pounces, or you see the bank of a river from a distance, before dying of thirst. Even though the fossilised remains of even older hominids have now been unearthed, dating from 4 million to 7 million years ago, it does not matter: for us, Lucy will always be the first and the most famous.

But let us not dwell too long on palaeoanthropology, a science as fascinating as it is difficult. Suffice it to say that, generation after generation, the grandchildren of Lucy grew and evolved in body and mind, and for about 3 million years, they lived in Africa.

In these 3 million years, or more, these strange new terrestrial bipeds, our ancestors, become stronger and improve their diet, and thus, their body, especially the skull, greatly increases in volume. All the while they stay in the same part of the globe, which indeed seems to contrast greatly with the behaviour of their later descendants. *Australopithecus afarensis* (Lucy is generally thought to be one of these) does not yet seem to want to explore.

It will take more than a hundred thousand generations, each with possibilities for more or less random genetic mutations, to change that attitude. And who knows how things really went: on such timescales, the truth of such things is difficult to assess. Although we cannot explain how and why this evolution took place—for example, whether it was gradual or occurred in leaps and bounds—what really interests us is that at some point, there came out of all the "grandchildren" of Lucy one who had a genetic heritage only a little different from that of Lucy herself, from whom also derives incontestably. It is this "mutant", and her descendants, who will be called *Homo sapiens*.

The difficulty of the explanation lies in the fact that, perhaps partly in parallel with the evolution of the entire African lineage of hominids descended from Lucy, there seems to unfold in Eurasia another evolutionary line of the genus *Homo*: lateral branches such as the Neanderthals and several others. However, in the end, *sapiens*, though a little jeeky and bully, wins out and emerges from Africa around 70,000 years ago to become a conqueror.

But there is more: there is a strong genetic link that is in common to all *Homo sapiens*. It is a small part—little more than a millionth of our DNA: 16,500 units out of more than 3 billions. This is our mitochondrial DNA: a small fraction of our genetic message, which is transmitted only down the maternal line. And, even up to and including today, all the mitochondria of our species come from one woman.

This mitochondrion African Eve is of course unknown, but she is truly the common "mother" to all of us that are now on planet Earth. She lived in Africa around 143,000 years ago, a date which can be given with some precision, since its current error is only 10 %. We are truly one big family, we terrestrial humans.

Why did *sapiens* succeed where other hominids failed? Maybe because of a new "organisation" of the skull: its volume of 1½ l is not flattened antero-posteriorly, as it is in the Neanderthals, but is much more spherical and with the bones rearranged, —evidently a morphology adapted to the development, first of thinking, and then, soon after, of language—the secret to the transmission of an individual's experience to all the rest of her species.

Or perhaps she/he succeeded because she/he possessed something deep down, a compelling push to move, to explore, to try something new, even if it involves enduring hardships, undertaking risks and suffering incredible losses, and all without looking back. Certainly something new, something vastly new had happened. After 3 million years of a hard life and marginal existence, all more or less in the same region of Africa, a species is born that will not stand still, never again. And one which continues to improve at a simply unimaginable rate, evolving at a pace never before seen in hominids until then.

"What do you say, Jules?" I ask my French Virgil "Have I made it too simple? I have condensed decades of scholarly and heated discussions, which are still going on amongst anthropologists today, into a few lines, in order to summarise the history of human evolution up to the threshold of today, to the people of the Anthropocene. "Mais non, mon cher" replied Jules, "It was, however, already quite clear to me; you know that I was a great admirer of Darwin ... ."

Chapter 2
Sphere 0 (or the Earth)

Homo sapiens' Conquest of the Land Surface

It is a fact that within a few tens of thousands of years after leaving Africa, that is about 70,000 years ago, *sapiens* came to supersede all other species in the genus *Homo*, driving them to extinction, but not before cohabiting for a while and acquiring a bit of genetic material, let's say that 2 % of Neanderthal that is in all of us and which can still occasionally leap out rather unexpectedly.

And so *sapiens* begin to explore and populate the entire surface of the Earth, what we call here "Sphere 0". He is already an artist and a thinker 18,000 years ago when the small *Homo floresiensis*, the last remaining species, other than sapiens, of the genus *Homo,* extinguished. And 15,000 years after the disappearance of the last Neanderthal, we find *sapiens* celebrating the fact that that he was the only remaining representative of the genus *Homo* left on Earth by painting the caves at Lascaux. We can find there drawings of bears, bulls and horses, but also one injured man and some things mysterious that we shall never understand.

The conquest of the Earth's surface took place at an amazing speed. Eurasia and Oceania were almost completely occupied within a few tens of thousands of years, that is by 30,000 years ago. The Americas from Alaska to the southernmost tip of Chile were colonized in a time span of perhaps 2000 years, from 16,000 to 14,000 years ago. All this took place at the incredibly high average speed of more than a kilometre per year.

An extraordinary thing about the speed of this expansion is that it was, in fact, an expansion, not a migration. Migration as we understand it today is, for example, more akin to that of the Lombards in Italy. Perhaps more correctly, migration is a cyclical phenomenon as exemplified by the African wildebeest, or eels, or salmon, where populations travelled for reasons linked to reproduction or climatic conditions or other reason not yet understood.

With *Homo sapiens,* this expansion was truly a large-scale phenomenon. It does not refer to the transfer of a particular people; rather, it was a spread or the expansion of a range seeking for new areas to inhabit. It was an expansion

© Springer International Publishing Switzerland 2015
G.F. Bignami, *The Mystery of the Seven Spheres*,
DOI 10.1007/978-3-319-17004-6_2

especially "out of Africa", which occurred perhaps several times in response to climatic variations and some of their consequences such as the changing levels of the oceans.

What drove *sapiens* to expand their range and seek new horizons? I do not know how to give a definite answer to this phenomenon. However, what we do know is that *sapiens* alone, among his many relatives within the genus *Homo*, undertook this global expansion, which was, in short, the exploration and conquest of the world. Where there were no predecessors, such as in Australia and the Americas, *sapiens* could easily settle without competitors. In other areas where related species were already present, such as in Europe, Asia and in Africa itself, *sapiens'* potential competitors soon faced extinction. Today, ethologists and ecologists agree on a technical definition that classifies *Homo sapiens* as an "invasive cosmopolitan" species. It is not really a compliment: it is a definition that applies equally well to rats or cockroaches, as well as to us, i.e. the only species that since the dawn of history appeared to have dominated the world.

The expansionist ambitions of *sapiens* may have genetic reasons: a variant (7R) of a gene called DRD4, which is responsible for the controlling of dopamine, a substance essential in the functioning of brain. The DRD4-7R variant seems to be present in about 20 % of all *sapiens*, and it might be the substance that gives us the impetus to look for new sensations, new in all senses, including food or sex, but also pushes us to take all the risks involved.

However, the discoverer of the 7R variant is himself the first to warn us about simplistic interpretations.

If it is true that 20 % of humans are inclined to risk and are in continuous search for novelty (including, e.g. cheating on spouses...) and that this could explain the rapid and violent expansion of a new species, it is also true, as Kenneth Kidd of Yale University says, that a single gene, or a variant of a gene, is not a sufficient cause.

It is equally true that several complex factors should to be taken into account: the genetic ones responsible for physical changes, as well as the environmental ones, and more. In short, motivation alone is not enough (although necessary): occasions are needed as well.

And so we arrive, about 10,000 years ago, at the end of the long phase of the Palaeolithic. The beginning of the Neolithic sees that *sapiens,* well established in all continents, the inventor and master of languages, able to produce and gather food and thus no longer forced only to hunt—in short, ready to move into the future.

And so the age of prehistory ends in an ideal fashion with Oetzi the Iceman, found preserved in an Alpine glacier at an altitude of more than 3,000 mt. He was well equipped to venture into the mountains with his leather shoes lined with hay to keep his feet warm, gloves, a hood, etc. He died from being shot with an arrow—perhaps while exploring he found himself in someone else's territory. Who knows? The detail is not now important since *sapiens* the explorer has won and in a flash has populated the whole Earth. The rest, as they say, is history, because it is from here on in that history begins.

"My Dear Jules, what do you say? Did you know the story of *sapiens the explorer*?"

"No, I admit that I did not know this—but—well—I had imagined some of it. In fact, if I could, we should now write a book... how about 'Around the World in Eighty Thousand Years'?" he replied.

"It would be one of your greatest hits... You know, shortly after your death we tried reading some of an American author's work, a certain Jack London who wrote "Before Adam" (1907), which is full of adventures and primitive men with sabre-toothed tigers; but alas, he had not your pen."

The history of the recent exploration of Sphere 0 begins, symbolically, with Oetzi the Iceman. However, it is a difficult story to tell simply because the greater, the bravest and intelligent explorers are unknown to us and will probably remain so forever. History from our "Western" perspective generally tends to present the geographical discoveries and explorations as all having been due to white men and to be related in some way to the formation and spreading of the great empires: for example, those early empires born in the Mediterranean, such as the Egyptian, Persian, Macedonian empires, and later the Romans, whose maps were annotated "hinc sunt leones", just lions beyond these boundaries, or to the later oceanic and colonial empires of, for example, Spain, Portugal, and Britain.

Schools today teach us that Asia was "explored" by Marco Polo and that America was "discovered" by Columbus and so on. Yet few in Europe (or at least outside China), know of the great admiral Zheng He, one of the most important explorers in history. From the year, we call 1405 under the reign of the second emperor of the Ming Dynasty (but which was, of course, a different date for Zheng He) over a period of more than 30 years, Admiral He explored throughly the Indian Ocean, navigated the entire coast of India and explored the east coast of Africa from Somalia as far south as Zanzibar. Zheng He was equipped with magnificent oceanic-class ships, although not as manoeuvrable as the caravels of Columbus a century later. Some argue (especially in China) that Zheng He rounded the Cape of Good Hope and then crossed the Atlantic to discover America... but experts say there is no evidence for this and that his ships could not have managed such a voyage. However, the Chinese could certainly have gone up the west coast of Africa and then end up in Spain, thus discovering and, why not, colonizing Europe... Thinking upon the real possibilities of such a voyage opens up fascinating scenarios of history fiction.

From the sixteenth century to the eighteenth century, after the great explorations across the oceans during the establishment of the Spanish, Portuguese, and British empires, and thanks to the Dutch trade routes too, we come to the great continental explorations of the nineteenth century, such as those in Africa by Henry Morton Stanley (1841–1904), the journalist–adventurer–explorer. While he did not find the source of the Nile, he did find David Livingstone (1813–1873) whom he had been rather vaguely instructed to look for as he was then "missing" "somewhere in Africa". The quest ended in that famous meeting between the two of them, the only white people for thousands of miles, when Stanley is reputed to have said,

"Dr. Livingstone, I presume?" They then proceeded to "explore" the course of the River Congo, led, naturally, by the "natives" who already knew the way.

Another interesting case is that of the exploration of the North American continent at the beginning of the nineteenth century. Between 1804 and 1806, Meriwether Lewis and William Clarke "explored" from the Missouri to the Pacific, where "no one" had yet arrived. They were acting on the instruction of the great American president Thomas Jefferson, (whose face is on the now endangered $2 bill), who wanted the newly formed Union to establish a trade route to Asia. But they did not go alone: they had with them an Indian girl of fifteen, Sacajawea, who was already a mother and the wife of an ambiguous French trapper. She knew the way as well as the various languages of the Indian populations that had long inhabited the area. She suggested the most natural passage through the Rockies, the Bozeman Pass (Interstate Highway 90 now passes through it), since she mysteriously knew that buffalo herds migrated through that pass.

Many of the great explorations that have been recorded and celebrated in our history, such as the circumnavigation of the globe by Magellan's fleet from 1519 to 1522, were in fact required to complete and possibly confirm the politico-religious vision that was, from time to time, prevalent in the Western world. Such explorations, however, were also pursued for commercial purposes: for example, finding the best route to the Spice Islands in 1500 promised yields on invested capital that amounted to profits comparable to those arising today from trafficking street drugs. The Spanish and Portuguese empires, in fighting each other over the Spice Islands, "discovered" that Africa and South America were not connected to Antarctica (which no one knew at the time), but that the southern tips of these continents could be navigated around, albeit with some danger and difficulty. After all, Africa and South America could have been joined to Antarctica, with just a slight change in the drifting of the continents, or a minor change in the form of Pangea, that ancient place where all the continents of today were combined in a single landmass before they began to separate and spread at about the speed at which fingernails grow. On second thought, for the same reason, the American continent might not have existed to divide the Atlantic and Pacific. Forget *"buscar el levante por el poniente"*; Columbus and his ridiculous caravels would have disappeared into the blue of a single Ocean joining Europe and Asia, just as the wise geographers of Queen Isabella had always said, since for them the Americas did not exist.

There were also embarrassing, even if unintentional, "rediscoveries". In 1512, Magellan, before being the first to pass through the strait that was later named after him, had been heading out of the Indian Ocean when he found himself in the archipelago we now call the Philippines. Then in 1521, coming from the Pacific, he rediscovered them without realizing it (and eventually even died, there, in a minor skirmish with natives onshore).

The islands finally took their current name when they were again "newly discovered", 20 years later, by Ruy Lopez de Villalobos during the reign of the son of Charles V, precisely Philip II. However, it should be said in everyone's defence that the Philippines are an archipelago comprising over 7000 islands, scattered over almost $2,000$ km^2 of ocean.

Of all the wonderful stories of those great explorations, one that perhaps stands out above all others is recorded in the diary of Antonio Pigafetta, the Italian nobleman whom Magellan took with him as secretary–interpreter during his trip around the world. Although loyal to the end to his Commanding General, he was mainly interested in the new geographical, environmental and anthropological disciplines. He noted, with passion and with a lot of imagination, such things as the vital statistics of the inhabitants of Patagonia, (who according to him were the tallest of peoples) or some seemingly impossible characteristics of penguins and seals. He even began to write a multilingual dictionary simply by pointing to known objects and writing in phonetic Spanish the names the natives gave in their own language. In short, he was a refined intellectual, but also one who was at ease at sea in the turmoil of a caravel of the year 1500, on which he sailed for months.

But when Pigafetta meets the Patagonians, one must ask: who explores whom? After all, the Patagonians had already arrived safely in Patagonia, perhaps driven by an earlier "Magellan" who remains unknown to them or us, and moreover, they went on foot, starting from the Bering Strait, or even farther afield. The same holds true for the sources of the Nile, which were well known to generations of the native inhabitants long before Livingstone tried to get there or to the hills of Montana, which the Sioux had long since explored and where they had lived undisturbed until General Custer arrived.

In short, the greatest explorers were those who first came to rest their eyes on landscapes that no *sapiens* had seen before. These people will never be known to us, but they certainly existed. Perhaps they were less individualistic than Christopher Columbus or Louis Antoine de Bougainville, who went around the world in 1769 and brought to European gardens the beautiful Bougainvillea flowery plant; or James Cook, who, like Magellan, circumnavigated the globe and was also killed by the natives on a Pacific beach; or of Alexander von Humbolt, who, as a scientist–explorer, studied and understood the physical geography and hydrology of South America.

"Our" history, however, does not say who led the Viking ships to the shores of North America, or who first explored and populated Easter Island and Hawaii in the Pacific between 3000 BCE and 800 CE, when they were really "discovered" for the first time. The ancient Europeans frequently explored lands that were already known to others. They did so for reasons of commercial expansion, sometimes so nefarious that they were often disguised as missions to spread religious belief. But what drove the ancient inhabitants of Taiwan in 3000 BCE to undertake long and dangerous sea voyages in fragile boats and so spread their culture and their language across the immensity of the Pacific?

It was probably exploration for expansion, simply driven by the need for living space. For people living on an island, with its clearly defined boundary, any increase in population would imply a depletion of scarce resources, such as agricultural land, and a consequent decline in the quality of life. Future Polynesians did not put a canoe in the ocean to go in search of gold or to convert someone else to their own beliefs, but did so for the most basic need of a growing population: to search out new land to live in and for their descendants to continue to inhabit.

The great continental and oceanic explorations of our Western history culminated in the late 1800s, the century of Jules and his book The Children of Captain Grant. In this beautiful novel—in which a Scottish Lord with the help of Paganel, a slightly potty geographer, takes around the world Captain Grant's children and a few other companions looking for their father—Verne uses a beautiful literary ruse. To justify travelling around the world from continent to continent, always in exotic, unexplored places, Verne uses a variant of the classic "message in a bottle" theme, which gives the story plenty of suspense. In this case, the message is found in the stomach of a shark. The message, apparently written in three languages by Captain Grant, asks for help and provides the coordinates of their position. But seawater and shark stomach juices have partially deleted the content of the message. Throughout the book, Paganel tries to interpret the erasures and guess what the missing data might be, but he is always wrong. Thus, Verne takes us around the world, forever chasing a new interpretation of the text.

By the end of the nineteenth century, there were few places left to explore on our Sphere 0. In reality, only those truly difficult or inaccessible places, such as the polar ice caps and the highest mountain ranges, remained unexplored. In the following years, up until the first half of the twentieth century, the last part of our history of exploration takes place. It is a fascinating period, because the drive to explore now comes from a desire for knowledge and from an individual's or country's striving to be the first to achieve the particular hard-won goal; that is, exploration for pride and glory rather than for practical or commercial purposes and its concomitant material rewards.

The race to the North and South poles was certainly a spectacular example. It became a sort of obsession for individuals and countries, including *fin-de-siècle* France old Verne who, of course, wrote a novel about the exploration of the North Pole in 1864: "*Les Aventures du Capitain Hatteras*". The hero of the novel, Hatteras, is an Englishman, who, with a handful of faithful companions, finally and after many vicissitudes comes close to the Pole. One of their parties is a doctor who knows all the answers, including how to make fire from ice. Right on the geographic pole, they discover an island with an active volcano. Its crater is in the exact position of the Pole, and Hatteras does not hesitate to go down it. At this point in the first version of the novel, Verne contrives an heroic death for Hatteras, but his tyrannical publisher, Hetzel, instead insists on imposing an ending in which Hatteras survives but loses his mind. Back in England, Hatteras is condemned always to walk only northwards.

In 1909, after numerous expeditions from different countries, the real North Pole in the middle of the Arctic Ocean was reached by an engineer in the US Navy, Robert E. Peary. Today, there remains some doubt as to whether he reached the exact Pole; but maybe it does not matter since Peary reached the Pole mainly due to a large company of Inuit who went with him, and who had, perhaps, already been there before without realizing it nor giving the location any particular importance. Certainly, they knew how to handle dogs, sleds, boats and how to survive. As always, even on the inhospitable shores of the Arctic Ocean, *sapiens* were happily settled, having long since become adapted to the conditions and having learnt how to survive in the most difficult of the ecological niches that remained free.

But the real challenge in the early twentieth century was the South Pole. There were no native Inuit there to act as guides, and it was not possible to get there by boat. This point of the Earth's axis is to be found on a plateau at 2,800 mt. of altitude, in the middle of a continent twice the size of Australia, visible from a distance and yet totally unexplored at the time. The South Pole could only be reached on foot, crossing glaciers and mountain ranges with skis, crampons, snowshoes and sleds. In 1911, two competing attempts were made: a British team led by Robert Falcon Scott, and a Norwegian team led by Roald Amundsen.

This was exploration in the purest sense of the word, very similar to that of space. No one had ever before been to the interior of the Antarctic continent, and no relief or rescue was realistically possible in the event of a disaster, nor of course communication of any kind. The Norwegians won the race, as is well known, in the sense that they reached the pole first and moreover returned alive. It was a fantastically impressive feat, helped by the fact that they were excellent cross-country skiers and they knew how to manage their dog sledges. Moreover, as some dogs became unnecessary as the load decreased, they had the courage to kill them for food (for the surviving dogs, but who knows…). The British led by Scott did reach the South Pole soon after Amundsen, but all died of hunger and cold on the way back. They had no dogs (which they would never have killed, of course), and thus, they were slow, since they had to pull the sledges themselves. They were also mediocre skiers and they miscalculated their supplies. Even heroes can get it wrong.

By a curious coincidence, in the years 1926–1928, both poles saw a new kind of exploration that Jules Verne would have been madly excited about. In 1926, the Italian Air Force general, Umberto Nobile, managed a spectacular achievement. He had designed and had built the airship *Norge*, a project which was paid for by Roald Amundsen himself, who was in fact on board when it departed from Rome's Ciampino airport. It headed north and flew over Europe, Norway and the Svalbard Islands and then northwards, flying over the Pole on May 12. They finally landed in Alaska, after an uninterrupted flight of 5,300 km in about three days. Two years later, Nobile attempted to repeat the journey, this time from Milan with another airship, *Italia*. Although the airship had reached the Pole, it crashed on the return journey. The subsequent agony on the ice, with casualties, some fatal, was a tragedy that affected Nobile, the great explorer, for the rest of his life.

In 1928, Admiral Richard E. Byrd of the US Navy established a base on the coast of the Antarctic continent in preparation for a memorable attempt to fly, on 29 November 1929 on a three-engined Ford plane in less than one day to the South Pole and back. To get to the Pole at more than 3,000 m above sea level, he had to lighten the plane, by throwing everything out—even the emergency food supply. Had Verne's imagination invented the stories of Nobile and Byrd, nobody would have believed them.

Verne certainly would also never have dared to imagine the following incredible feat, which falls halfway between exploration and an endurance test. In 1989–1990, Reinhold Messner and Arved Fuchs crossed the entire Antarctic continent in

92 days, on skis and with no outside help. In what was perhaps the greatest and most authentic feat of exploration undertaken in Antarctica since that of Amundsen in 1911, they made only one stop at the US Polar Base to celebrate the New Year.

In some sense, mountaineering is symmetrical and similar to polar exploration, especially those climbing expeditions that involve some element of research and exploration. Such adventures among mountain ranges began during the eighteenth-century enlightenment, continued into the romantic nineteenth century and culminated in the troubled twentieth century. Within Europe, the summit of Mont Blanc was, for centuries if not millennia, in clear view to those in the fully explored and civilized areas of Italy, France and Switzerland, yet its summit was only reached in 1786. During the next century, especially thanks to English mountaineers, all the great peaks of the Alps were reached and then the great mountain ranges of the other continents of Africa, Asia, the Americas and Antarctica—almost all less than a century ago, although such achievements were often gained by explorers and climbers only with support from local guides.

Symbolically, the exploration of the Earth's surface, the 'Sphere 0', ends on 31 May 1953 when Edmund Hillary and Tenzing Norgay conquered Everest and Hillary says: "Above us, the sky". The news, and the phrase, arrived in London just in time for the coronation of Queen Elizabeth II, who still reigns at the time of writing this.

It seems that, in the mountains, each of us can be part of a real process of exploration, not only when we find a new route up a mountain, but also simply by putting our hands on a rock that nobody has ever touched before. Even when I climb a mountain along a route that someone else has climbed before, I put myself voluntarily and for no apparent practical reason in the path of the unknown and of danger—why on earth do I do it?—I ask myself. Sometimes, deliberately and yet with a certain level of detachment, one heads out to a climb simply by looking at the mountain, without following someone else's route, as if one were the first man (or woman) in the world to do so. In short, there are always new ways to be discovered, and it is enough simply to have the desire to do so.

When Walter Bonatti, who was one of the greatest mountaineers in history, realized that he was no longer able to "write new pages in the mountains" as he put it, meaning that he would no longer be able to establish new routes or climb virgin peaks, he became an explorer of distant places, about which little or nothing was known. A spectacular proof of the link between mountaineering and exploration.

In the 1920s, an American journalist (who understood little about mountaineering) asked George Mallory, why he wanted at all costs to climb Mount Everest (where he eventually died), Mallory answered: "Because it's there". Such an elegant reply went down in history. In the 1950s, another great climber, the Frenchman Lionel Terray, who in 1952 conquered and explored mountains like Mount Fitz Roy in wildest depths of Patagonia, admirably summed up the spirit of alpinism in the title of his famous book that, once again, would be the envy of Jules Verne: "*Les Conquérants de l'inutile*". These conquerors of the useless are well exemplified today by the twenty climbers (a group whose number is growing rapidly) who have reached the summit of all 14 mountains with peaks higher than 8,000 mt.

It is a very small club whose first member was, of course, Reinhold Messner, in 1986, and which today is growing in a similar way to the number of astronauts who landed on the moon or have otherwise left Earth's gravitational pull (currently 27, as we will see later).

It is not only the mountains but the sea too, which has tested the true nature of *sapiens'* modern taste for exploration. For example, in 1947, a group of anthropologists led by Thor Heyrdahl set off from Callao, Peru, on the Kon Tiki—a raft of balsa wood—and successfully navigated across the Pacific to the Tuamotu Islands in French Polynesia. They wanted to show that the Polynesians' ancestors could have done so thousands of years before and so populated the Pacific. It is not clear if the Kon Tiki crew were right, but the adventure would almost be even more beautiful if it were now proven to have been "useless". The Kon Tiki expedition has now been beautifully echoed, more than 60 years later: Alex Bellini, an Italian mountaineer from Valtellina, rowed alone across the Pacific Ocean, again starting from Callao. He arrived in Sydney, Australia, in 2008. If it was not madness, it was certainly pure exploration.

Alex certainly does not only have an outstanding physique, but also a good quantity of the 7R variant of the DRD4 gene, the one that pushes you towards the unknown no matter the conditions.

Another important event in the modern history of exploration, which is also splendidly useless, falls halfway between Sphere 0 and Sphere −1 (the exploration of the oceans). It is the entirely submarine circumnavigation of the globe, following the route of Magellan. This was completed by the nuclear submarine, *Triton,* of the US Navy, in 1960. The undertaking must be considered in a sense to be almost a sport, since it is an end in itself. In fact, it was accomplishment during the Cold War and thus was kept secret by the US Navy. Leaving from the secret base of New London, Connecticut, the *Triton* was first brought to its ideal starting block, represented by the St. Peter and St. Paul Rocks, which are remote, uninhabited rocks off the coast of Brazil, and which were first sighted by Magellan in 1520.

On February 24, the *Triton*, always submerged, took a route to the south-west, arriving off Cape Horn on March 7. The *Triton* then had to divert, rather than navigate submerged, as might have been wished for historical completeness, through the strait found bravely by Magellan in 1520. The official reason for this single deviation was that the Magellan Strait is in Chilean territorial waters and asking for a permission to the Chilean authorities would have compromised the secrecy of the mission. The real reason, we say maliciously, and without any foundation, was that to plumb the tortuous depths of the Magellan Strait, which were certainly not charted in detail, would have been perhaps too much for the submarine's commander to contemplate, since a US nuclear submarine run aground or sunk in the Strait of Magellan would have not made great propaganda for the US Navy.

Our heroic divers entered the Pacific and went on to Easter Island, which Magellan had missed. From the periscope, they could see and photograph the mysterious stone statues, just to prove that they really had been there. It was then on to the Philippines, approaching dangerously the coast of the island of Mactan, on

whose shore Magellan was killed by the natives. In another rare moment at periscope depth, the commander permitted the officers on the bridge to see through the
eyepiece the monument to Magellan erected on the beach. By coincidence, they
also saw the only human being spotted throughout the mission: an unknown
Filipino quietly paddling his canoe. History does not record whether he saw the
periscope, and if so, what story he then told at home. Then the Nautilus sailed on to
Borneo, Indonesia, the Indian Ocean, the Cape of Good Hope and back to St. Peter
and St. Paul, where they arrived on April 10.

In total, the *Triton* had travelled 26,723 miles, always submerged, and had
circumnavigated the world in 60 days and 21 h on the route which took Magellan
nearly three years to complete.

"What have you got to say about that, Jules?" I ask my friend, "Impressive,
although Captain Nemo with his *Nautilus* in *"Twenty Thousand Leagues Under the
Sea"* is still unbeaten (2000 leagues correspond to 60,000 nautical miles)", and
Jules continued *"Eh oui, mon amis, ces americains...* less than a century after my
book. And in reality; not in fantasy! I admire them very much, *pas question*. But,
maybe they got a little inspiration from me" And I am happy to say this to him: "Of
course, Jules, the first American nuclear submarine, in 1954, was named *Nautilus*,
in great honour of Captain Nemo. Viceversa the *Triton* has also gone around the
world quicker than your Phileas Fogg in "Around the world in eighty days" "Sure,"
Jules replied, "but you want to compare the poetry of my balloon with that iron pipe
of the Americans, always under the water?" "Well, today, you can fly around the
world in an aeroplane in less than 80 h, if you want to; and, if you keep on reading,
soon you will see that by staying just above the atmosphere, it is achievable in little
more than 80 min".

How many people have lived on Earth from the time since the first *Homo
sapiens* appeared, about 200,000 years ago? Seven billion people are currently
living, but since the beginning of *sapiens'* adventure, (according to a necessary
rough estimate) 107 billion men and women have lived on Earth, all within less
than 10,000 generations. Is this too few or too many? Is it a significant figure in the
evolution of life on our planet? Absolutely not, if we compare it to the numbers of
individuals of those species that were around for hundreds of millions of years—
species such as the trilobites or even the dinosaurs whose demise allowed the rise of
the first mammals, since the Tyrannosaurus was no longer around to eat with gusto
our very small, warm-blooded, hairy, marmot-like mammalian ancestors.

Or, to put it in another way,—if the total human population today were spread
like jam on the surface of the Earth, how thick would that layer be? Actually, it
would be about the same as if 2.5 kg of jam were spread uniformly over a square
kilometre, i.e. 0.0025 mm/m^2. And even if we multiply that figure by fifteen, to
include all those 107 billion *sapiens* who have ever lived, it still would not be very
much thicker, we would always be invisible.

Another way, more romantic but even more significant, to see the role of human
species, not only on our planet but in the context of the whole Universe, is as
follows. The one hundred billion (10^{11} in scientific notation, i.e. 10 followed by 11
zeros) of successive sapiens on Earth have been and are made of "stardust", that is

"baryonic" matter, matter with a mass, as everybody now knows. But what fraction of stella matter in *sapiens*? Easy: together, we humans represent one part in 10^{41} (10 followed by 41 zeros) of this type of matter in the universe.

It is a very, very small fraction. How small? Well, to give an idea let us say that in all the Oceans of the Earth there are 10^{46} water molecules and that we humans compared to the universe are like a hundred thousand molecules of water, that is an infinitely small and totally invisible fraction of a drop.

Too hard? Let us try instead to think that this infinitely small fraction of the matter in the universe, over a time scale of less than a millionth of the life of the universe in which it lives, is able to understand such universe, its size, its content of matter and energy, its evolution and much more.

Nearer to us, the energy, *sapiens* have used in 150,000 years that have passed while forming our history, has mainly been the metabolic energy of muscle power (i.e. biochemical energy). Later came the discovery that the combination of certain elements, such as carbon, with oxygen in the air, above a certain temperature is violently exothermic, giving off heat: that is to say, it warms you. *H. sapiens* did not know that he lived immersed in a sea of oxygen or that wood is made of carbon, but once he discovered fire, learning that mammoth steaks are better grilled than raw was a simple experimental observation. As the story unfolds, an increasing number of energy sources are used, all available directly from Mother Nature. As well as wood and coal, there are wind mills and sails, the sun to dry meat and figs, the water of rivers for mills and transport, and the metabolic biochemical energy of domesticated animals.

Much later, i.e. from a few centuries ago to the present day, we, the modern version of *sapiens*, have used and still mainly use the thermal energy derived from fossil fuels. It is a small step then, with respect to the grilled mammoth, to move forward to those other great discoveries by *sapiens*—the electric pile by Alessandro Volta in 1800, and nuclear energy by Enrico Fermi in 1942.

Remember, however, that burning a gram of gasoline releases about 11 kcal, that is the highest concentration of energy per unit mass of chemical material that exists, except for that released in the formation of the water molecule. It is even higher than chocolate or butter, although these are much tastier than petrol. Of course, in some cases, the chemical energy contained within gunpowder, or the electrochemical energy within batteries, has come to the support of *sapiens'* explorations. However, it also came in handy to settle other little internal affairs of that warlike roughneck whom *sapiens* has always been, culminating in the end, of course, with the use of nuclear energy.

Chapter 3
Sphere −1 (or of the Oceans)

Exploration of the Ocean

In 1953, the same year that Everest was conquered, Auguste Piccard and his son Jacques made a record dive with the bathyscaphe *Trieste* to a depth of −3150 mt. in the sea off the coast of Ponza, an Italian island off Naples. A few years later, in 1958, the first nuclear-powered submarine of the US Navy, *Nautilus*, made its way to the arctic, reached the North Pole and surfaced off Greenland. Thus began the era of ocean exploration from within, or from below, rather than from above. Then, two years later, in 1960 (the same year *Triton* completed its submerged circumnavigation of the globe), the *Trieste,* with Jacques Piccard again on board, and with US naval support, reached a depth of 10,916 mt.: the deepest point of all the Earth oceans, at the bottom of the Mariana Trench. It is just one more courageous example of what it means to explore: this time it is the depths of the ocean and its floor, which no one then had yet seen; and even now, those few who have seen it have only seen a small part.

Although the sea makes up nearly ¾ of the Earth's surface, undersea exploration has only just begun. Unfortunately, even in this case, as it has been for the exploration of much of Sphere 0 (except maybe for the summit of Everest), it has so far mainly been driven by the pursuit of profit. In addition to searching for sunken Spanish galleons and their treasures, such exploration is primarily oriented towards the exploitation of oil and gas—fossil fuels beneath the seabed. Something has been discovered almost everywhere, such as off the coast of Gela, in the Mediterranean, in the North Sea and in the legendary Gulf of Maracaibo, where Emilio Salgari, who needed no oil, once roamed in the *Thunderbolt* of the Black Corsair, hunting down and hopefully sinking Spanish galleons.

The ocean floor is unexplored because it is difficult to access; but it is covered in interesting things—not just the wreck of the *Titanic* or the *Andrea Doria* and thousands of other ships of the past. For example, all the meteorites that have fallen into the sea over billions of years have had a softer landing and are quite well preserved—certainly better than on land. A simple calculation will give us a rough estimate to illustrate the point. The current rate at which extraterrestrial matter falls on the Earth, in the form of dust, water, metal and rock, etc., is about of

© Springer International Publishing Switzerland 2015
G.F. Bignami, *The Mystery of the Seven Spheres*,
DOI 10.1007/978-3-319-17004-6_3

40,000 tonnes per year. In the first billion years of the Earth's existence (which is 4.65 billion years old), the rain of meteors and comets was even greater—much greater; so much, for example, that it seems that icy comets were largely responsible for forming the Oceans.

The total amount of material that has fallen on the Earth is therefore more than 40,000 × 4.65 billion tonnes; perhaps one and a half times that amount. In short: a lot. In fact, the extraterrestrial material-turned-terrestrial is so much that, if we were to imagine it as being spread evenly over the entire spherical surface of the Earth (= $4\pi r^2$, where r is the radius of the Earth = 6,300 km), it would form a continuous and uniform layer about 1.5 mt. high—all over the Earth (maybe a little less, let us say one metre, due to the presence of water in meteorites). We are not aware of living on a layer of meteorites because the surface of the landmass is continually being altered by meteorological phenomena, volcanic activity, tectonic movements, etc., and now, even by anthropogenic influences.

An interesting example is that of gold. Recent studies have shown that all the gold that we know of, and which has gone into coins, or that we wear around our necks or on our fingers—or in our teeth—in short, everything that has been mined from the Klondike, for example, is largely of meteoric origin. It seems that the gold that we find in mines fell to Earth long after the formation of the planet and arrived on board these meteorites.

It is easy to calculate why this is so. Due to its specific weight, which is about four to five times that of terrestrial rocks, all the primordial gold that must have existed on Earth since its formation from the proto-planetary disc slowly fell towards the centre of the Earth before it began to solidify.

Returning to the idea of a layer of meteorites over the entire surface of the Earth, we should remember that, due to the extent of the oceans, three-quarters of all meteors fall into the sea; the effect of their impact and their subsequent preservation is quite different, depending on whether they fall onto the sea or land. On the sea, there is a "splashdown", which is much softer than an impact with the ground. The meteorite slowly sinks for perhaps a few kilometres until it comes to rest on the bottom, where many meteorites, especially the small ones, arrive basically intact.

As for their conservation, one should consider the volcanic and tectonic activity occurring on the seabed, as well as the motions of currents and the deposition of sediment; but the overall effect on extraterrestrial material, at least over the course of a few hundred million years, is much less severe for objects landing in the sea than for those landing on the surface of the Earth (which is also much more populated).

So, there are great opportunities to explore millions and millions of square kilometres of the ocean floor in search of extraterrestrial rocks, perhaps of Martian origin, gold nuggets, or perhaps simply iron almost ready for use (to wit the famous ferrous gigantic meteorite in Greenland, from which eskimos crafted spearheads for centuries...).

Nobody yet knows how to explore this marine environment without adversely interfering with it, i.e. without ruining the plankton or scaring the whales. Above all, no one yet knows how to create a political body that could have the overall management and control of such undersea operations.

Perhaps such control would be even more important for another global submarine activity, this one already in progress: the exploration of the mid-ocean ridges and nearby volcanoes, where many of the tectonic plates meet and diverge, and of the nearby "ring of fire", where the continental plates meet the oceanic ones.

Until just a few years ago, many were sceptical about the possibility of mining the ocean floor, and even drew similarities between mining operations on the sea floor and on the Moon, which is notoriously unprofitable. But, in the last few years, attitudes have changed: the reason being that land-based mines are being impoverished of rare minerals, as well as of the ever-important gold. It is now becoming an attractive proposition to think about mining the most interesting places on submarine ridges, where the volcanic deposits are most abundant. The technology used for prospecting and mining underwater is now being rapidly developed and uses robotics under the control of surface ships.

The hunt for submarine sulphides has already begun in the Pacific Ocean and in the Indian Ocean, with strong participation from countries such as China, Japan and South Korea. But even private American companies are exploring the beautiful waters between Fiji, the Solomon Islands and New Guinea.

But how can one "explore" the very depths of the oceans? Large-scale mapping has already been done; but the detailed exploration of those areas considered to be particularly interesting will require vehicles that have yet to be invented. We know they must clearly be cheaper and more effective than existing submersibles and underwater robots, since all are operating only at "short range", not least due to the lack of visibility at such depths.

One also needs to understand that, in this case, exploration, as has always been the case in human history, also means exploitation. Two very large problems immediately arise, which have not yet been neither resolved nor dealt with. The first, macroscopic, is ethical and environmental: the bottom of the sea belongs to everyone, starting with the animals that live there, from plankton to whales, with their complex biological food chains. And first on the list to be protected are those strange and unfamiliar forms of life that inhabit the space immediately surrounding submarine volcanic fumaroles, or critters happily thriving in extremely hostile environments.

But it is precisely there that the most interesting minerals accumulate, sometimes in spectacular chimneys and steep mounds, all made of products dissolved in the fumarole vapours and then condensed and precipitated abruptly in the surrounding icy seawater. One of the few known deposits of this type has been appropriately named *Godzilla*—a name suitable for a 50 mt. tower of sulphides. There is no point in pretending that any exploitation of the Oceans' minerals will not interfere, in an unpredictable but potentially disastrous way, with marine life in that same environment where, it would be well for us to remember, life started on Earth.

The second problem, no less important than the first and related to it, is political. Outside of the coastal zones of territorial waters, the ocean floor is for everyone, or anyone, and all should be respected. Someone must, for example, regulate the concessions granted for the exploration, prospecting and exploitation of all kinds. So far everything has been entrusted to a sleepy agency of the United Nations, located in Jamaica—the International Seabed Authority. Recently, the ISA has been submerged, it must be said, beneath a flood of questions concerning the exploitation of sulphide deposits and the like. Unfortunately, the ISA may not have the necessary authority and ability to respond with the appropriate speed to control a phenomenon that is currently undergoing such a dangerous rate of expansion.

We would need the rapid intervention of the United Nations, firstly to impose regulations that will safeguard the environment, and then to establish the potential ownership of the supranational seabed. Such a legal framework should already have been put in place some years ago when, in 2007, an ex-Soviet nuclear submarine planted a Russian flag on the sea floor of the Arctic Ocean, well outside Russia's territorial waters, and in an area naturally rich in oil and gas. And this is very reminiscent of what the Soviets attempted to do in 1959 when, with Luna 2, they launched on the Moon some metal plates with the emblems of the USSR.

Just like the Americans did with their flag planted by Armstrong and Aldrin ten years later.

"Dear Jules, a few years ago a clumsy modern imitator of yours, Michael Crichton, described in detail an underwater city …" A grim look from Jules: "How dare he! And what's more … *un americain?*" "Yes, Jules," I say, "In his book *Sphere*, published in 1987, he describes in detail the technicalities and the construction of an underwater city, comprising a group of towers, with its entrances and exits hundreds of metres deep. Actually, Jules, it's not bad. There is some science fiction, and some mystery and suspense; you should read it …" "*Jamais!*"

Is it a purely sci-fi idea? It would certainly be a useful thing to accomplish in order to help in the detailed exploration of the seabed. We may try, with a minimum of realism, to imagine how such a base might look like, and what living in it might be like for *Homo sapiens oceanicus*.

The starting point is the pressure that prevails at such depths and its effect on the human body. In particular, the effects of that most dangerous of all the gases that make up our atmosphere—oxygen, which is a very aggressive gas precisely because it is an oxidant. The respiratory system of *Homo sapiens* has evolved to breathe oxygen at a certain concentration—about 20 %—as it occurs in that gas mixture which we call the atmosphere. Recall that much of the rest of the atmosphere is made of the chemically very inert gas—nitrogen. Our bodies can handle these gases very well at the pressures experienced where we normally live.

Going deeper, things get complicated at the higher pressures that occur there. In apnoeic diving, that is, simply holding one's breath while underwater, *Homo sapiens* has made incredible progress in the space of less than a century. The depth record for an apnoeic dive under "constant weight", i.e. diving unaided to gain depth and up again, using only the strength of one's body, just as the seals do for example, is already approaching 130 mt., and continues to increase. In the "variable weight" technique, i.e. using weights to gain depth more rapidly, some divers are now approaching the incredible depth of nearly 300 mt.—a challenge even for a seal. These depths are achieved all within the 2–3 min duration of a "dynamic" apnoeic dive (the record for a static apnoea is more than 18 min). These time scales make it difficult to speak of exploration.

Things get more complicated if we also want to breathe, in order to remain under water long enough for at least a look around. Meanwhile, we might say that one cannot think of exploring the bottom of the sea using a tube or pipe like a long snorkel from the mouth to the surface (as kids, myself included, often use to think). The water pressure on the chest wall at a depth of, for example, 10 mt. is 1 kg per cm^2 for a total weight of 5,000 kg. And for the lungs to be in equilibrium with the atmospheric pressure through a pipe would be impossible as the chest would be exposed to a pressure equivalent to having a large truck parked on it. Death would follow rapidly, as the chest wall would collapse. No, one needs to find another way to breathe and survive in order to explore the depths of the sea.

Everyone knows that the easiest method is to breathe properly compressed air, so that the pressure inside and outside the lungs is equalized. It is a method that has been used successfully by divers for some time, with no problems to depths down to a few tens of metres. At depths below that, once again further complications set in due to the laws of physics concerning the solubility of gases in liquids. Our blood does just that, among other things; i.e. it transports oxygen around the body to all the tissues that need it, the oxygen being bound to haemoglobin in an elegant gas exchange process in the lungs.

However, at a pressure of about three atmospheres, the nitrogen present in the compressed air dissolves in the blood, something which does not happen at normal pressure. If one returns to the surface too quickly, a physical phenomenon—similar to what happens when a bottle of champagne is uncorked—occurs in the circulatory system: bubbles form which are carried around in the blood, where they cause dangerous and life-threatening embolisms. To avoid this happening, one must undergo gradual decompression to allow time for the nitrogen to come out of solution and be expelled in the breath without forming bubbles.

At even greater depths, perhaps below 100 mt., where it is frequently dark because water, even though clear, absorbs sunlight, the fun of exploration really begins; but a human being can no longer use ordinary compressed air to get there. In fact, if breathed at a pressure of ten atmospheres, which is required at a 100 mt. of depth, the oxygen content of normal air is too high for the human body to deal with, as oxygen becomes toxic and quickly oxidizes lung tissues, that is, it "burns'" them. Instead, a different "atmosphere" is needed, one that has been invented ad hoc during the development of undersea exploration. Instead of nitrogen, another more inert gas is used: helium, which does not give problems of bubble formation. Oxygen is added to helium at a much lower concentration than exists in ordinary atmospheric air. The result, however, provides the right amount of oxygen by supplying less of it, but at higher pressure, so avoiding any lung damage.

Using this type of breathing gas, together with special equipment to protect against the cold and to provide sufficient lighting, a human being can now free-dive to depths of more than 300 mt., something unimaginable in the times of Verne. It is likely that within a few years, the depth record will increase further, driven by scientific, commercial and military interests in this field.

Let us imagine a realistic limit to be 500 mt. for a human to comfortably explore the sea depths. We would then see that, if the 500-mt. isobath around the coast is drawn on a map, a significant percentage of the oceans (for example, those around the North Sea, the China Sea, part of the Arctic Ocean and many other areas) would become accessible.

But would these be accessible only to the best equipped and most highly trained supermen, sponsored by navies or research centres? Certainly they would at first. But remember that, less than a century ago, it was almost impossible for anyone to venture even a few metres under water; and even just a few decades ago it was still a rare activity reserved for reckless eccentrics. Today, with ever more sophisticated equipment and a wealth of experience, many are able to dive to considerable depths because of breathing apparatus. An army of amateur explorers have already found in coastal waters plenty of interesting things hidden in the sea, such as the Riace bronzes. There are also, among these, those who think they have found Atlantis … but explorations inevitably include among their number a few hoaxes. Or are they?

But in the case of Atlantis, this fascinating hypothesis has a certain credibility according to the geo-archaeologist and underwater explorer, Jacques Collina-Girard. Plato's dialogues, Timeo and Critias, are our best sources on the legend of Atlantis. In the first, Plato makes only a vague reference to the island of Atlantis and its demise. In the second, a more complicated and discussed work, partly perhaps apocryphal, Plato tells of a great civilization of Atlantis, located on an island beyond the columns of Hercules, in an opposite direction to ancient Athens. Is it pure fantasy or is it a literary ruse to quibble about governments and social organization and to show that the Athenians were always the best ones? The answer is in another dialogue on the "ideal government", *The Republic*.

What is certain is that the legends of Atlantis have their foundation in these references to Plato. Was Plato himself telling about ancient Mediterranean stories? Assuming that Plato, who lived about 2,500 years ago, had heard stories passed down through the millennia, Jacques Collina-Girard, the French scientist, conducted a serious study on how the landmass was distributed at the end of the last ice age, between 19,000 and 12,000 years ago. There is no doubt that the sea level was much lower at the time, because of the presence of huge water masses frozen in glaciers lying inland, or "solid water" that covered not only most of Europe, but also other continents. During the glacial period, it is known, and has been verified, that the sea level was between 200 and 50 mt. lower than it is today. A lot of ocean water was essentially out of the oceans.

The coastlines of both Europe and Africa, when drawn with the water reduced to the −200 mt. isobath, appear very different from what they are today. In particular, there is no doubt that on the Atlantic side of the Strait of Gibraltar, between Cape Spartel and Cape Trafalgar in Spain and Morocco, various islands emerge. These include one particularly large island, Spartel, and some that would be clearly visible from the "Pillars of Hercules" and surrounding areas. Here, perhaps, is the source of the legend that Plato had taken from Mediterranean stories, passed down over thousands of years by oral tradition and handed on to us. Who is to know?

What is certain is that it would probably be worth exploring the unnamed island submerged at −52 m off the coast of the Strait of Gibraltar (as big as the Mediterranean Pantelleria Island). One should not do it not in the hope of admiring the Atlantis columns and palaces that Captain Nemo used to show his guests from the great underwater windows of the *Nautilus*. Rather, more realistically, it would make sense to undertake a palaeo-archaeological exploration for traces of Neolithic artefacts, which are hard enough to find on the dry land of the Earth's surface. To do this, today seems a hopeless task, but tomorrow—maybe not. Unfortunately the result would not be as revolutionary as one might think: yes, we may find that, on a piece of land that is now under water but which was dry 12,000 years ago or so, there were the same sorts of settlements for which evidence is also found today on the Spanish and Moroccan coasts. It would be a sad end to the myth of Atlantis.

But the real professional activities of underwater exploration cannot be limited, either in depth or breadth. Similarly, in the more or less near future, we will inevitably have to overcome the limitations linked to the resources of the human body. As already mentioned, explorers would need a basic refuge and living space, which would obviously have to be pressurized. For structural and access convenience, we might imagine them in the same bathymetric region where man can (or will be able to) move freely, but which could be built and maintained by robot submarines controlled by a few humans in a bathyscaphe. Perhaps, our friend Michael Crichton was right in his fascinating book "Sphere". A permanent submarine base at medium pressures, say about 500 m deep, would be crucial for the exploration of the surrounding seabed, especially from the scientific point of view.

For example, the wealth of underwater life at a certain depths remains largely unexplored. There are some basic problems concerning the physiology of all life in the ocean that it would be worthwhile studying; knowledge of which might be of great advantage to human physiology. For example, some marine species, such as shrimps or octopus, use copper instead of iron (as we do) for the transport of oxygen in the blood: instead of haemoglobin they use haemocyanin, which gives the distinctive blue colour to these species' blood. Copper is much more efficient than iron in transporting oxygen to tissues, thus it allows life to exist where oxygen is in short supply.

Haemocyanin-based blood is not only blue, but is also much denser than our own red blood, and so requires a greater effort to pump it around a body. That is why some marine animals, for example, the very common octopus, have a large heart, or actually more than one. This story of haemocyanin-based blood also provides a small but brilliant bit of additional evidence on the marine origin of life on earth: scorpions, for example, whose form resembles that of a small lobster, have blue blood.

There are also some medically useful chemicals that have been extracted from the blue blood of a living species, but one that should be a fossil because it has been around for hundreds of millions of years, namely the horseshoe crabs of the Atlantic coast of the USA. I wonder, however, what else might arise from serious research on the physiology of other marine species that are relatively "alien" to us. From marine biological research at serious depths, we may gain potentially therapeutic benefits, especially because a lot of "alien" underwater life remains unknown and awaits our discovery.

One very promising dimension to the sphere of the abyss is the production of food for the population of the Earth. The demographic explosion of *Homo sapiens* on land has created a problem concerning the future production of nutriments. The land itself, which basically covers just less than a third of the planet's surface, cannot all be used easily or to its full capacity from the point of view of the agriculture. Moreover, those areas that are accessible are often badly used, sometimes resulting in a heavy environmental impact. Today, in our quest for food, we mostly exploit the sea's producer of protein and oils, i.e. fish; but few eat, for example, algae. Daily consumed in Japan and in Asia, algae are still lacking on Western plates.

We know that seawater contains, even though at very dilute concentrations, a large amount of microscopic life, both animal zooplankton and vegetable phytoplankton, which by photosynthesis of sunlight forms the base of many food chains. In short, exploration and research may teach us how to feed ourselves on the rich, if diluted, fish and vegetable soup that is the ocean. Similarly, we might imagine growing crops directly on the ocean floor itself, at least down to the depth reached by light, but perhaps even beyond that. The "soil" of the ocean floor is not exploited from the agricultural point of view, although it is certainly rich in minerals and organic matter: an ideal place to grow algae genetically designed to be palatable and of high nutritional value. Will we finally have an underwater potato? Perhaps, though we might need to cross it with algae.

But there is more. Nutrients, such as organic minerals, periodically well up from the depths of the oceans to form a type of local springtime at the surface, a phenomenon which differs in its timing in the two hemispheres. The enrichment of surface waters by minerals from deeper water thus periodically generates an increase of plankton—those micro-organisms at the base of the ocean food chains. It is an interesting example of the interaction between the mineral content of the largely unexplored ocean depths, and the abysmal life, or simply the marine life, from which, let us remind ourselves again, we and all other forms of terrestrial life are direct descendants.

To highlight the interaction between mineral nutrients and ocean life, it is worthwhile mentioning another curious phenomenon of land–sea interaction, which is also more or less regular, but this time occurs at the surface. In the Northern winter season, the Harmattan, a strong wind, blows from east to west over the Sahara creating large sandstorms. Thanks to satellite images it is now easy to see how the fine sand of the Sahara is carried by the east wind to great distances into the Atlantic Ocean, where it is deposited over huge areas of the sea, up to the America coast.

This shower of sand on the sea's surface creates areas where the seawater is made correspondingly richer in mineral nutrients. This is where we observe, also from satellites, a significant increase in the surface plankton density and, consequently, of all the marine fauna.

Thus, there are some mineral nutrients that well up from the sea floor, and some which derive from the sand and other material that falls onto the sea's surface; but it does not matter from where the nutrients originate, since in either case the result is the production of life, from the microscopic scale to the macroscopic scale, as the chain of all life has its beginnings in the minerals of the Earth's crust.

It is a phenomenon that, although it still requires a lot to be understood in detail, could be essential in helping a planet that will certainly be more and more hungry in the future. Clearly, the exploration of the seabed and, more generally, an increased understanding of the role played by ocean minerals, both those contained within the waters and those always falling on them, may be vital for a future *Homo sapiens "oceanicus"*.

The methodical exploration of the fauna of the deep, such as might be possible from a submarine base, could also help us understanding the evolution of life on Earth, for example in its crucial stage of transition from water to land. That this took place we now know with certainty, but why it happened and how it happened we still do not know. Was it a spontaneous expansion, or a random process, or was the tremendous selective pressure within the water enough to push fish and shellfish to try their fortunes out on dry land, in a vast ecological niche that was yet to be inhabited? The answer to this and many other fundamental questions about the details of biological evolution would greatly benefit from the systematic exploration of life underwater.

Another area that still needs to be explored more fully, and which is most important to life and found most abundantly in the oceans, is the presence of the water itself. We think we know a lot about the water of the oceans, and it is true we certainly do; but we have not yet figured out, for example, how much of it is of extraterrestrial origin—that is, how much of it has fallen to the Earth when it was being bombarded by comets and asteroids, a process which occurred mainly in the first billion years of the planet's life, but which still continues today. Nevertheless, it would be ultimately interesting to know, when we swim in our seas, whether or not we swim in an extraterrestrial liquid. It is also logical to ask whether the ice of comets contained organic materials or even life itself, when they impacted on the Earth. The answer is YES to the first question, and DON'T KNOW to the second.

In the field of practical applications, however, we still need to take advantage of the numerous metallic and non-metallic elements dissolved in seawater, and to extract them from it. We might start with gold, perhaps, but continue with the "rare-earths". It is a group of little-known and rare elements, but which are important for many current and future high-tech applications. These elements are already rare on Earth and are becoming increasingly rarer and, perhaps due to their uneven distribution on the surface of the Earth, are considered to be the "property" of a nation (e.g. China, in this case). If we assume that the sea will remain up for grabs by anyone for a long time, and if we can find ways of extracting these elements from the water in ways that will not damage the sea itself, then even this seems to us something that needs to be explored further.

In the field of submarine bases, before Crichton's work of the imagination, a small but real base has already been established. One of the fathers of undersea exploration, together with Auguste Piccard, was another Frenchman, Jacques-Yves Cousteau. He did not use submersibles for deep-water exploration, but was among the first to scuba dive with portable air bottles. He also made, among other things, important contributions to research on the physiology of human respiration at high pressure.

Among his many wonderful initiatives and enterprises of exploration, there was one that was, above all, romantic. He started building a small underwater "city" in the Red Sea in 1963, which has now been abandoned for many years. It comprised two or three rooms, at various depths, where one could live for many days in a row, breathing, in a pioneering way, a helium–oxygen mixture. One got in and out by using scuba diving gear, but when "in residence" one could dress normally and eat at a table watching sharks through the window.

Cousteau did his research in his underwater home in the Red Sea. But I wonder what Jules Verne would say today, were he to know that in Key Largo, Florida, there now exists an underwater hotel bearing his name: "*Jules' Undersea Lodge*". It is pressurized, and one enters by swimming into it. It is full of rooms and has a dining room with portholes from which to observe underwater life. And who knows, maybe the manager calls himself as Captain Nemo—let's hope not.

Chapter 4
Sphere −2 (or of the Nether World)

Journey to the Centre of the Earth

In Jules Verne's *Journey to the Centre of the Earth* (1864) its protagonist, Professor Otto Lidenbrock seemed to find everything easy. In an old book, he had discovered a loose sheet of paper, written in ancient runic by the Arne Saknussemm, in which the great Scandinavian explorer of the past claimed (in Latin) to have been to the centre of the Earth. It seemed simple enough to follow his instructions. A particular Icelandic volcano, the Sneffels Jökull, has in its crater three deep shafts. The central one, says Saknussemm, goes directly to the centre of the Earth; just go down it. Verne's intrepid explorers throw themselves into a series of breath-taking adventures and end up floating in a wooden raft on a sea of lava (an adventure which they miraculously survive) and are "erupted" back to the surface—surprisingly still alive-out from the mouth of Mt. Stromboli. Unfortunately, it was nothing like the centre of the Earth; nevertheless, it was a great subterranean adventure, which made us all dream.

Today, in the field of human exploration of the Earth's interior, we are far ahead of Professor Lidenbrock. At least we know that an understanding of the interior of our planet is more important than ever, even for what we hope will be the long survival of *sapiens*. So far, we have explored and continue to explore underground here and there, but we have only scratched the thinnest surface skin of the Earth, say up to a depth of one-thousandth of its radius—only 6 km of its total radius of more than 6,000 km. The fact that the record depth to which a well has been drilled is nearly 20 km does not change the situation.

In the best tradition of modern human exploration, we went underground especially in pursuit of profits: mining metals, diamonds and coal and then drilling wells for oil, natural gas, etc. But we have learned a lot from this direct exploration by man, in particular we have been able to assess the available fossil energy reserves present on Earth and which have been derived from the early plant and animal life on our planet. In relation to the energy needs of humans and to the thermal characteristics of our planet, in particular to the increase in temperature observed at greater and greater depths, let us advance a modest proposal, which

© Springer International Publishing Switzerland 2015
G.F. Bignami, *The Mystery of the Seven Spheres*,
DOI 10.1007/978-3-319-17004-6_4

could be one of the first practical results of the exploration of the depths of the Earth.

We can start from the basic principle that the Earth is a ball with a hot interior. Below the lithosphere (110 km), there are the mantle, which is about 3,000 km thick and made of hot rock, and the core, 3,500 km thick, which is very hot, between 3,000 and 6,000 °C, so that heavy metals contained therein are partly in a molten state, despite the very high internal pressures. Heavy metals "sank to the bottom" in the Earth's early life, when it was still a molten sphere: lighter rocks (the average density of the mantle being between 2.5 and 3) now float on the core, which has a much higher density, so that the average density of the Earth is 5.5. The reason for the Earth's high internal temperature is simply the sum of two factors: one is the original heat generated during the formation of the Earth, dating back 4.65 billion years; the other is the radioactive decay of elements, such as thorium and uranium, which are heavy and thus abundant in the Earth's core.

The hot core transfers its heat to the thick mantle of the rock, which fortunately is an excellent insulator that retains heat, but allows some part of it to be conducted near the surface. On average, over the whole Earth, the temperature of the mantle increases by more than 30 °C per km depth, although there is a considerable amount of local variation. Indeed, for many parts of the Earth, including Italy, the temperature at a depth of 5 km has already been mapped. It shows that in many parts of Italy, and even more so in Turkey, the temperature of the rock at depths of up to 5 or 6 km is between 200 and 300 °C. In short, there is a huge supply of energy, which is essentially infinite and free, ready and completely accessible right under our feet.

From the earliest times, mankind took advantage, and still does, of sources of geothermal energy located close to the surface. Italy pioneered the construction of the first geothermal power plants, followed by Iceland. Now, there is the possibility of doing something much more general and complete: to take advantage of the deep geothermal energy, which means to systematically extract heat from the rocks in the first few kilometres of the continental lithosphere, of course after surveying it to identify the most convenient places.

How heat is extracted from the rocks of the lithosphere is obvious: one uses a heat exchanger as simple and non-polluting as water is. First, to carry water down, a well is sunk to the required depth (5–10 km), which is then extended horizontally creating a cavity in the rock for the exchange of heat. At the end of the horizontal portion, another vertical borehole is sunk from the surface. Into it, vapour ascends at temperatures between 250 and 300 °C, depending on the depth of the boreholes and local conditions. This is more than enough to turn a turbine and directly generate power. There you have it: clean electricity, free and infinite for all. Too simple? Seeing is believing: many other countries are already doing it.

After this example of the practical use that can come from the exploration of the Earth, we now come to another aspect of scientific research in the deep subsurface: the discovery of life forms capable of living at incredible depths. Take, for example the strange, thermophilic and anaerobic bacteria which have adapted to living at high temperatures without oxygen, and which have been observed in a gold mine in Africa at depths greater than 3 km.

Searching for these extreme forms of life is difficult, especially owing to the danger of contamination by surface-living life connected with the excavation. In other words, one must be very sure that the probe that drills deep into the Earth to bring up samples to the surface has not first been in contact with normal life forms at the surface—for example, it must not have been left lying on the grass.

Having taken the proper precautions, the search for subterranean life is an absolutely fascinating project of scientific exploration. To what depth will we still be able to find such life? A few kilometres, as already observed, or much more? And, what temperatures can such a subterranean life withstand?

What can we learn from a systematic search for deep subterranean life forms? For example, it could turn out to be a very old form of life, already present since the time when life in the seas and on the surface of the Earth began to create the oxygen that has since been present in our atmosphere .... And if it is really that old, deep underground life (and who knows how deep) could it not even be the first form of life on the planet? It is difficult to assess the pros and cons of the arguments, but maybe deep down, towards the centre of our Earth we might find an answer to the quest for our origins.

This would make an exploration of the deepest bowels of the planet rather like travelling back in time. Who knows what secrets, even perhaps about the origins of life, are hidden in the depths of the Earth, at least as far down as where temperatures do not reach prohibitively high values? We might be inventing the new science of geobiophysics: that, alone, justifies making the effort to go deeper and deeper.

In his story, when Verne was pushing Professor Lidenbrock towards the "Centre of the Earth", as the depth increased it was like, for him, a travel back in time. At one point, he passed through a sort of gallery of the oldest fossils ever. And then, on the shores of a huge underground lake, he encounters two dinosaurs fighting each other, and many species of mammals now extinct, even a kind of human ... Verne's imagination roams freely away over this, more so than over anything else.

But the search for other forms of life within the Earth could have a very interesting corollary in the search for life on other planets of the solar system, most notably Mars. After all, there is no doubt that the solar system planets were all born at the same time and that the four rocky inner planets are very similar. If life, at least in its initial and its most elementary state, was formed underground, maybe even very deep underground, might it not have appeared, and perhaps still be present, on Mars or Venus? On the surface of these two planets, today we know that conditions are very difficult, if not impossible for life. But deep down, in a warm, protected environment, safe from radiation and toxic gases—why not? After all, some forms of current life have learnt to eat stones, perhaps seasoned with a little oil, which many bacteria have been making from time immemorial.

Besides direct exploration by drilling wells and boreholes, how else can one explore the unexplored, which forms the vast, inaccessible majority of the volume of our planet? Well, we get a little help from lava, the molten rock that is brought to the surface by volcanoes and which acts as a potential messenger from the planet's interior, rather like meteorites are messengers from outer space. Indeed, the comparison between the two types of samples (terrestrial and extra-terrestrial) is particularly interesting if we take into account the common origin of all "rocks" of the Solar System, which are all derived from the nebula from which our Sun was also born and which has itself a very similar composition. Unfortunately, even the lavas are not, at least as far as we know, from the really deep zones of the planet (and maybe it is a fortune, as they would probably be highly radioactive). And so, even volcanic lavas give incomplete information, since their origin is from volumes of the Earth which are limited in depths, and, moreover, lavas are extruded only in volcanic zones.

So far, most exploration of the extreme depths has been indirect and has been carried out using seismic waves. They are waves that propagate in, and bounce off, the Earth's layers, like acoustic vibrations in a ringing bell, at least for those who can hear them.

Even earthquakes can serve science. Indeed, since seismic waves, which are normal mechanical waves, do not leave the planet, they can be heard rebounding around inside the Earth. The ways in which the various layers of the Earth respond to the mechanical stimulation of an earthquake (or any other source of vibration) are valuable for understanding the nature, form, extent and so on, of the various layers. But they also help, it must be said, in the discovery of oil fields or underground lakes. However, we are talking about depths of the order of a few dozen, or at most a few hundred kilometres, i.e. that part of the crust where all earthquakes occur and which is basically very small compared to the total volume of the planet.

Earthquakes, of course, are a serious potential threat to the natural environment, including people or objects that *Homo sapiens* has spread over the surface of the planet. But, above that, they are a spectacular manifestation of plate tectonics. The great internal heat of the Earth generates convective movements from the centre to the surface, much like boiling water in a pot for making pasta. On the surface, the convective thrust from the interior of the Earth moves the continental masses at speeds and directions that today have been very accurately measured—that is at a speed slightly slower than that at which human fingernails grow. But this convective motion from below is actually what helps to sustain life on Earth. Without a continuous supply of nutrients from deep within the Earth, the surface would, in fact, become sterile within a few tens of millions of years and life would end up starving for nutrients or might never have begun. A little evidence of this phenomenon lies in the quality of those plants that have their roots in lava. We all know how good are the oranges of Etna, or wine made from grapes grown on Vesuvius. It is a strange and terrible paradox of Mother Nature that one of the most important features of the Earth, the transport of nutrients to the surface, is also the cause of earthquakes that can be so devastating.

Of course, if it is used for geophysical surveying, instead of natural earthquakes, we can create man-made seismic waves. In the 1950s, some of the biggest thermonuclear bombs, with the power of tens of megatons of TNT, were tested without problems in underground cavities, which were built just for that purpose. It was after all, better than continuing to conduct explosions in the atmosphere. In the USA, you could easily see the effects of a seismic bomb exploded underground in Siberia; likewise, those in the USSR could "hear" what was being exploded in New Mexico. It was an expensive and dangerous way to take an image, a radiography of our planet. Even today, Norway has an excellent network of seismic detectors placed in the far north of the country along what was once the Iron Curtain. Today, there are no more nuclear explosions in Siberia, thankfully, and Norwegian seismographs have an important scientific role in an interesting a little-studied area— listening to the planets interior in a largely uninhabited region, which is therefore devoid of background seismic noise such as trains or trucks passing by.

The seismic method was also used during the Apollo programme in order to understand the structure of the Moon's interior. During the first missions to the Moon in 1969 and 1970, the astronauts left seismometers on the surface; the only thing missing was a tool to create the mechanical waves to produce the seismic waves to "X-ray" the Moon. With an aim to remedy this, subsequent missions did not let the last stage of the Saturn V rocket, which always accompanied each Apollo module leaving the Earth, escape into some diverging orbit. From Apollo 13, this last stage was carried to the Moon and then dropped at high speed onto its surface. The impact of several tons of metal, travelling at a few km per second, was enough to make it resonate so much so that, at first, it seemed that it might even be hollow inside. The seismographs left by the previous missions could measure in detail the effects of the impact. It was a good experiment, one from which we learned a lot about the internal structure of the Moon, but at a price, unfortunately, which was the pollution of the Moon's surface with numerous pieces of iron made in the USA.

The study of the deep internal structure of the Earth is needed, among other things, in order to increase our understanding of the origin of our magnetic field. Its position, structure, and variability are now quite well known, but its origin and in particular its future evolution remain largely unknown. To understand this really well, we need to realize Verne's dream, at least to get as close to the centre of the Earth as possible. Is it impossible? Of course, men such as Professor Lidenbrock may be able to do this; but perhaps it may also be possible for us mere mortals to explore the very depths of the Earth by using some ingenious and futuristic method. According to a well-known planetary scientist at Caltech in Pasadena, D. J. Stevenson, it is possible, even if it sounds incredible, to send a "satellite" to observe in situ the centre of the Earth. Not from a distance, and not from orbit: but actually in the centre of the Earth. The idea of a robotic mission to the centre of the Earth is based on principles similar to those of space exploration, but it is on a much smaller scale than the planetary distances. Or rather, it is reduced by a factor of one billion, which is more or less equal to the difference in energy required to travel every metre in depth. Flying in space is easy, once you leave the Earth. "Flying" within the interior of our planet to the centre of the Earth is much more difficult: to travel one metre below ground requires a billion times more energy than does in space. If one accepts that this enormous amount of energy is needed to descend into the depths of the Earth, one can imagine an underground satellite, a "space" capsule equipped to move into the underworld. It would be a ball of tungsten, the size of a grapefruit, able to send to the surface all the information gradually discovered. How would the information be sent? Simply, by using high-frequency seismic waves, which would be transmitted very easily and travel well and at high-speeds through the thousands of kilometres of rock. How would we receive the information? With a special kind of purpose-built seismograph, which is already well within our reach, at least conceptually.

How could we sink our probe down through the depths of the Earth? This is the difficult bit; but, mind you, it is not impossible. The small tungsten probe could be sunk within a flow of molten iron, which would be starting from the surface of the Earth. There it would be poured into a fissure, which would propagate itself downwards, stopping only at the centre of the Earth. The principle is similar, only in the opposite direction, to that of the ascent of magma, namely the migration of molten volcanic material through the lithosphere to the surface. Given the difference in density between iron and rock, getting the flow started would require, for example a 10 cm wide fissure in the rock, 300 m long, 300 m deep, and filled with 10,000 m^3 of molten iron (or a few tens of thousands of tons of iron, which is the product of one hour of all foundries in the world!). Such a cast would fall inexorably unstoppable at about 5 m/s. This is in fact the speed at which the fissure would propagate, if it had been opened with sufficient energy. This last question of opening the crack is the most critical point. The energy required would be quite high—the equivalent of a few megatons. We have more than that in our nuclear arsenals, which instead of sitting there rotting (or worse) could be put to better use. It is not a huge amount of energy: the total energy required is something like that of an earthquake of magnitude seven on the Richter scale.

The fissure would thus first be formed with a nuclear explosion, and the liquid iron would then be poured in together with our exploring-transmitting ball of tungsten, which of course would not melt in the molten iron. After the process has been triggered, the fissure advances downwards, little by little, closing above itself as it goes. The journey would continue for about a week, which is all the time required to reach the centre of the Earth. Once there, the ball, which would have on board miniaturized instruments yet to be invented, would be activated to measure such things as the temperature, electrical conductivity, chemical composition and whatever other parameter it might be possible to measure, and the data transmitted to the surface. For the first time, we would have an idea of what that mysterious and totally unexplored thing is really like—the centre of our planet, where a lot of important things happen that affect our lives.

For example, we still need to understand the history of the Earth's magnetic field. No one doubts that its origin is related to the planets core, molten on the outside, and high in iron and other magnetic elements. But no one knows why the magnetic field that aligns our compasses from the North Pole to the South Pole, "occasionally," about every 800,000 years, but quite irregularly, changes its polarity, thus swapping the north with the south. Do not panic: for us earthlings the transformation takes place on timescales of hundreds of thousands of years and has no immediate practical consequences. But it is a fascinating problem of geophysics, perhaps related to the differential rotation of the molten outer core and the rest of the solid planet.

We also know that an efficient and orderly magnetic field is essential for life on Earth. Mars, for example, does not have one, and so the influence of solar storms reach much nearer to its surface, where they would be inimical to life. Exploring the deep interior of our planet would also tell us where all the gold missing from the surface might be. Even more helpful would be our finding a means to preserve our own precious magnetic field, first of all by understanding, from measurements taken in situ, exactly how it evolves.

The robotic exploration of Sphere −2, the sphere of the Nether World, proves to be perhaps the most difficult, so far, of all areas of exploration. Verne was not even able to think about it very much. Yet, the principle of a subterranean "space" probe, initiated by nuclear energy and then driven by gravity, could potentially open up one of the greatest areas of scientific discovery. Moreover, under our very feet, there is our planet, our magnetic field; in short, things that affect our lives in so many ways, but about which we still know so little.

Chapter 5
Sphere +1 (or of the Sky)

The Conquest of Circumterrestrial Space

We have heard how on 31 May 1953 Edmund Hillary, standing on the summit of Everest, the last unexplored point of Sphere 0, said: "Above us, the sky". Eight years later, on 12 April 1961, there was a man in the sky, above the atmosphere, conquering the circumterrestrial region of Sphere +1: Yuri Gagarin, the first man in orbit.

From the 300–400 km altitude of his single orbit, Gagarin says (in the recordings you can hear his official, military voice become a bit excited; after all, he was only twenty-seven year old): "I see below me the Earth ... it is blue ... it is beautiful..." He is the first explorer to enter a new dimension, the first truly new dimension for mankind. For an hour and forty-eight minutes, he was above the planet, with its atmosphere, where *Homo sapiens* was born and had hitherto been confined.

Man's dream to explore space beyond the Earth's surface and also beyond the atmosphere (which for us forms a part of Sphere 0) has its roots in the ideas of one particular man, among many other prominent scholars from around the world: Konstantin Tsiolkovsky[1] (1857–1935). He was more or less a contemporary of Jules Verne, a fact he does not seem to have particularly appreciated or even known (of course he wrote in Russian, but he must have been fluent in French, like all Russian intellectuals at the time). Anyway, it was Tsiolkovsky who first thought of actual, physical, long-duration space flight. It was he, above all, who wrote "The Earth is the cradle of mankind, but a man is not made to live in a cradle". And to think that Tsiolkovsky did not know that Africa was the real cradle of mankind, as we now know it to be, nor that we had all migrated from it, a long time ago.

But who really was the first to open up Sphere +1? Who was the man behind Gagarin's flight, and who, before Gagarin, had invented Sputnik? It was a little-known character, indeed, a man who was for many years kept a strict secret by the obsessive Soviet regime. It was Sergei Korolyov (1907–1966) who had always been passionate about ballistics and missiles and as such was considered suspect by Stalin who, in the 1930s, had Korolyov interned in a gulag, one of the infamous Soviet prison camps. While there, he lost most of his teeth and contracted serious cardiopulmonary disorders which ultimately brought him to a premature death. But,

© Springer International Publishing Switzerland 2015
G.F. Bignami, *The Mystery of the Seven Spheres*,
DOI 10.1007/978-3-319-17004-6_5

while in the gulag, he also came to know Andrei Tupolev[2], the father of the Soviet air travel, who was also interned as a dangerous intellectual himself.

After some years, however, both Korolyov and Tupolev were deemed to be beneficial to Soviet Union's war effort and so they received slightly better treatment, although photographs then taken of Korolyov in his uniform show him to be remarkably slim compared to his naturally stout physique that came back a few years later. After the war, however, Stalin first and then Nikita Khrushchev, the new all-powerful party secretary, ordered a full-scale programme to be launched, aimed to the building of intercontinental ballistic missiles that could carry the new weapon, the atomic bomb, right into the house of the enemy, the very distant USA. Although Korolyov's identity was top secret, he was the undisputed leader of the Soviet missile programme. His identity was so secret that no one knew his name; at most one could identify him by his nickname: *balscioi straitel*, "The Great Constructor", and whoever needed to understand, would understand. Korolyov was the first to successfully launch a rocket with a range of 4,700 km, the R-7 (*Raketa siem*), the seventh model of a lucky rocket series. The R-7 became Korolyov's workhorse; he was so proud of it that he went as far as calling it *Semiorka*, which is a colloquial Russian name for a tiny child, born in its seventh month. Semiorka itself, of course, was the biggest rocket ever built at the time.

When the moment came to scale up even *Semiorka* to a size capable of transporting bombs to New York and Washington, Korolyov wanted to know one of the best kept secrets of the Soviet Union: how much the bomb weighed and how big it was. To stay on the safe side, the regime gave Korolyov a ridiculously high figure —pretending it was much heavier and bigger than it really was. The *balscioi straitel* went to work in his disciplined way and soon produced an even bigger *Semiorka,* capable of delivering a payload of several tons to the USA. With great joy, Khrushchev, who had succeeded Stalin after his death in 1953, announced it to the world.

But, in the meantime, Korolyov had done some maths: given the power of his new *Semiorka,* he calculated that, with it, he could put a satellite into a more or less stable orbit around the Earth, if only they would let him do it! In 1956, Korolyov asked Khrushchev's permission for putting in orbit the first Earth artificial satellite. After consultation with an indifferent Central Committee, Khrushchev gave the official permission, a bit reluctantly, but knowing that he could not deny Korolyov something that was, in his opinion, a useless scientific toy. The Great Constructor had, after all, done a lot for their country.

The rest is history: on 4 October 1957, Sputnik, containing only its batteries and a transmitter and weighing a total of 83.3 kg, became the first artificial satellite of our Earth. Rather, in the best Russian poetic meaning of the phrase, our first "travelling companion": "sputnik": "s" is the Russian preposition for "with", "put" is "to travel" and "-nik", of course, makes it all become a person. Khrushchev immediately understood that this meant something more than a scientists' prank. The whole world was looking up to the sky or listening to the now legendary "beep, beep" on their radios. The Americans were in a panic: Sputnik passed over the USA several times a day, and at night you could actually see it! A few weeks later,

Korolyov mastered another coup, by sending into orbit on November 3 a much larger and heavier satellite. This one comprised a real capsule with a passenger— the little dog Laika, who immediately became a celebrity all over the world. While in orbit, Laika survived for some time and all the world could listen her heartbeat.

With a mammal similar to a man having gone into orbit, it was clear what was going to happen next. In fact, a group of young Soviet Air Force officers had already been sent, in total secrecy, to a very special training course. None of them was very tall, because Korolyov knew he had to save on the volume of the capsule in order to transport humans, and the shortest, who was only 1.57 m tall, but very tough, was called Yuri Gagarin.

And what about the USA? The land of the Wright brothers, pioneers of the heavier-than-air flight and home to those who made the first powered flying objects in the first decades of the 1900s, even by the end of World War II, harboured no visionaries of extraterrestrial flight, such as Tsiolkovsky, nor a bulldog-minded engineers, such as Korolyov.

But victory in Europe in 1945 had brought home, among other things, a rich booty for man's future exploration of space. It was the majority of the scientific team from a village on an island in the Baltic Sea called Peenemunde, where the V1 and V2 (V is the initial letter of the German word Vergeltungswaffe—the "payback weapon") were made in order to "repay" the allies for their systematic destruction of German cities from the air. The V1 and V2 missiles, which were already very technologically advanced, especially the V2, were launched to terrorize England and Belgium occupied by the Allies. Some members of the Peenemunde team were taken by the Soviets, but the head of them all was a young German scientist, who managed to escape deportation to the East (or the history of space would have been quite different). His nobility was perhaps a bit doubtful: some say that the "von" was a later, unilateral addition to his surname, but for all he was Wernher von Braun (1912–1977), the man who would give us the Moon.

He was a visionary of extraterrestrial flight as well as a leading expert on ballistic missiles. In April 1945, he cleverly managed to be captured by the Americans, avoiding the Russians. He was taken together with a train full of V2s, replacement materials and fellow missile scientists and technicians. While all this was later being transported through the deserts of the south-western USA, the band had some problems. The administration of the US President Eisenhower, the man who had the crusade against the Nazis in Europe, was not too happy about using former enemies, who were under suspicion of war crimes, to work on a US military programme.

Von Braun, in particular, had a lot to be forgiven. But having been born in 1912, he could perhaps be partly excused since he was a member of that entire German generation who were only twenty when Hitler came to power. However, not only was he a convinced Nazi, but, after spending a few years in research as a civilian, he later became a senior SS officer in 1940 and took command of Peenemunde. The mere fact of belonging to the body of the SS was considered a war crime in the USA even for the rank and file, let alone for an officer, a Sturmbannführer, a major, even if his rank had been given to him, he said, as an honorary position. After a few years of hard research work, it eventually came out that in Peenemunde von Braun and his officers had widely used prisoners of war and political prisoners as slave labour. This also meant that he was accomplice in acts of serious cruelty, including executions. In the years immediately after the war, the Allies prosecuted and shot or hanged culprits who were guilty of much less.

But von Braun had brought with him a train full of V2s and was a recognized genius of rocket missiles. He was not put on trial—he was simply left in solitary confinement in New Mexico for a few years. For a young man of action like him, boredom was a big problem. He asked for pencil and paper and, in 1948–1949, he wrote straight off a cross between a novel and a treatise on astronautics, *Das Marsprojekt*. It is a fascinating book because it is complete with all the information needed to go to Mars. Wernher von Braun also showed some ability in describing how to manage an international crew, strictly composed of only of men and with a paramilitary structure, of course. In addition, the book shows that he had already understood perfectly the psychology of politics and of the US military at the time. From this, it appears that he must have read *From the Earth to the Moon*, in which Verne devotes much attention to US politics and its interaction with major space exploration projects.

Von Braun and his men, after finishing several tests on the V2s brought from his homeland, began to design and test new, ever larger and more powerful models— much more powerful models as part of a US Army programme. Because it was the Army that had captured von Braun, the Army believed that they "owned" him; moreover, von Braun also faithfully believed that he belonged to the Army. However, in the mind of the Pentagon, it was the Air Force that was to deal with rockets and consequently much time was lost in useless rivalry.

Von Braun sizzled: he knew very well that they and the Soviets, and in particular his rival colleague Korolyov, were preparing for the same thing. He tried to say that his Jupiter rocket, even though it was an Army missile, was ready and could very well put a satellite into orbit, and well before the Russians could do so. Nothing to be done: he had to sit back and laugh bitterly while the US Air Force met with one embarrassing failure after another. After Sputnik, however, his time finally came: President Eisenhower called him (well, he called von Braun's boss, since he did not speak with Nazis …) and von Braun promised that he could get an American satellite into orbit within one hundred days—following, of course, the Army pro-gramme. After some understandable hesitation by the Pentagon, Eisenhower gave the go-ahead for the programme to start. And von Braun kept his promise: on 31 January 1958, Explorer 1, the first US satellite, was in orbit.

It was the beginning of the space race; the time when we, the *Homo sapiens* species, started to explore and occupy the circumterrestrial orbits of Sphere +1. The US had the technological lead, and important scientific results were quickly obtained thanks to the likes of James Van Allen, discoverer of the circumterrestrial radiation belts, and also of Italian immigrant geniuses such as Bruno Rossi and Riccardo Giacconi. NASA was soon established, in 1958, as a government agency dedicated to all kinds of civilian space activity and especially in order to meet head-on the Russians, who seemed so hard to beat, at least initially.

In the same year, 1958, an important body of global significance, and one which was requested by the United Nations, was also founded: The Committee on Space Research or COSPAR for short. Any nation that does, or is willing to do, space research can join COSPAR, which now numbers 46 states among its membership. During the years of the Cold War, it represented the only hope and forum for dialogue between space scientists of the two blocks of the East and West. It now continues to work towards building a politically international forum for space research, rigorously with civilian and non-profit aims; the formation of such a body is the only real step that needs to be taken for our planet to advance deeply into space exploration.

But to return to the newly formed NASA in 1958: it was formally headed by a civilian, but in reality completely ruled by the military, who maintained a separate space programme for the various armed forces, particularly the Army, which was adored by von Braun. This was a weakness for the USA, compared to the USSR, but it is an unavoidable price for democracy. While the programme of scientific satellites continued with great success in Sphere +1, where there was yet much to be discovered, von Braun obtained permission for his team to begin a programme of manned spaceflight, using the Jupiter and its even more beautiful and powerful derivatives.

Unlike Korolyov, NASA and von Braun proceeded more cautiously, and so, perhaps inevitably, lost time while performing sub-orbital tests—i.e. sending a capsule, the famous Mercury, up to an altitude of about 200 km—still with a speed well below the 8 km/s required to enter into orbit. Having reached beyond our atmosphere for a few minutes, and fleetingly also into the realms of zero gravity, the Mercury capsule would be left to fall back to Earth. As in the case of Sputnik, however, only a few weeks after Gagarin came the USA's response.

On 5 May 1961, Alan Shepard became the first American to clear the atmosphere in a 15-minute sub-orbital flight. Pumped up by NASA as the first American "astronaut", he was snubbed by the Russians, who had sent their beloved Juri into a real orbit. But behind NASA there was von Braun, with his organization and its ability to produce rockets at will; the Mercury Project is still seen today as a model of efficiency. Within a few months, John Glenn became the first American in a real orbit; after him, there were many others, all within less than two years. At first, of course, all American astronauts were men. It is a story seldom told how much women were regarded as second class—in politics as in everything else—in the USA during the 1950s and 1960s. When selecting the astronauts for the Mercury Project, a number of women, all highly qualified pilots, had come forward. History tells us that NASA did consider their applications, but when the time of the real choice came, President Lyndon Johnson seems to have intervened in person to say "no" to women astronauts, worried about the "social disorder" that would follow.

However, the "socialist paradise" of the USSR was shrewder about these things and their propaganda value: in 1963, the Soviet Air Force's Second Lieutenant Valentina Tereskova became the first woman in space. The first American woman, Sally Ride, only went into orbit twenty years later.

There began, between the two superpowers, a race to be the first to achieve various records in Sphere +1: first were the individual missions of more and more successive orbits, then with double or triple vehicles, then the first spacewalk (which requires the difficult and delicate job of conceiving, producing and developing a suitable space-suit) and then the first hook-ups in orbit or dockings of different vehicles. The more or less explicitly stated goal, however, was the race to the Moon—the conquest and exploration of Sphere +2, which would be unattainable for a few years yet.

But there was not long to wait: NASA's soon passed from the Mercury programme to the Gemini programme, based on capsules with a crew of two people, and then to that of Apollo, with three men on board. We will soon see, talking about Sphere +2, that Apollo would be the decisive programme, run in an incredibly short time by modern standards, and at a pace that the Russians were not able to match, in part because of the untimely death of Korolyov, in 1966.

Jules Verne had already planned, and admirably described, a trip to and around the Moon, but he did not foresee the technical and scientific achievements of socialism. The Americans won, dear Jules, but we will tell this story later.

After its phase of lunar missions, the Apollo programme finally ended with the last human presence on the Moon in 1972. So it was back to Sphere +1, also to dispose of the stock of those Saturn V rockets, the wonder rocket of von Braun, who survived a brutal cutting of the programme. After some semi-serious attempts at *detente* in orbit with the joint Apollo-Soyuz missions, in which American and Russian astronauts flew happily together, NASA started a new programme in 1975. They had a grand design that in theory was planned to enable astronauts to come and go to and from the orbit of Sphere +1, more or less like a round trip from home to work by bus.

It was called the Space Shuttle programme. The Shuttle itself made its maiden flight in 1981, twenty years after Gagarin's first orbit. The Shuttle was designed to carry into orbit, piece by piece, a permanent structure, a space station whose contents and functions were then still a little vague. However, building a large space station, permanently lived-in and usable as a base from which to explore and monitor the Earth, proved to be very hard, much harder than foreseen. To start with, it required a vehicle that could come and go quickly, at low cost (in theory) and have the capacity to carry a large payload, both human (six astronauts) and material (18 tons in the hold). In short, the Shuttle was designed to do just this.

On the other hand, once the lunar programme was finished and vanished along with von Braun (who in 1972, at the end of the early phase of the lunar missions of the Apollo project, abruptly ended any relationship with NASA and later died in 1977), also the possibility of a nuclear rocket to Mars, which von Braun had planned in detail, had to be shelved. It was in the same years, from the end of the 1970s to the 1980s, that the idea of a large space station was conceived jointly by the US Department of Defence and NASA. It was the apex of the Cold War, when the two superpowers were running strictly independent space programmes and were quite hostile to each other. A US space station was intended to bury forever the rival

Soviets under the weight of US technology, or so President Ronald Reagan's hawks thought, and they were later proven to have been at least partially right.

Reagan had even given the name *"Freedom"* to the space station still to be built, of course on an orbit inclined at 28° to the equator, which is the latitude of Cape Canaveral, the launch site of the Shuttle. But NASA soon realized that they had before them a *Mission Impossible,* if they were to try to achieve this with their own resources. During the 1980s and early 1990s, the *Freedom* remained a (very expensive) paper project, and nothing ever came of the permanent orbit at 28°.

Meanwhile, the Soviet comrades had got their station into orbit. It was a "no frills" job, as pragmatic as the rest of their space programme. Since 1986, the Mir space station had been in orbit, having evolved from the previous seven very spartan Salyut stations that had been sent into orbit since 1971. The Mir (which in Russian means both "peace" and "world") was launched by a Proton, the jewel of Korolyov's rockets, which as a carrier is still in use today, and a worthy successor to the *Semiorka.* It was launched from the same cosmodrome as Sputnik and Gagarin, i.e. Baikonur in Kazakhstan, at a latitude of more than 46° north. Launches to support and manage the Mir continued until the end of 1991, when the last cosmonauts who went up as Soviets and had to return to Earth as Russians. The programme was a success, but, as well as speaking Russian, it was in an orbit inclined at 51.6°. As we shall see, this orbit is very different from those preferred by the Americans, and it is very ill suited for departing towards interplanetary exploration.

At this point, not only had the Berlin Wall fallen, but the whole world had also undergone a major change. At last, NASA seemed to be really able to work with the Russians, something which for years they had more or less secretly dreamed of doing. Politicians jumped at the wonderful opportunity presented by the new situation. In 1993, the vice-presidents of the two superpowers at the time, Al Gore and Viktor Chernomyrdin, signed an agreement to jointly build a large space station, to be called the International Space Station (ISS).

Irrespective of the political result, came the first technical problems. To change from the inclination of one orbit to another, you have to spend energy, that is to burn more fuel, maybe much more fuel than that required to gain the original departure orbit. Moreover, celestial mechanics (in particular, the sense of rotation of the Earth) dictates that it is easier to increase the inclination of the orbit, for example, from 28° to 51°, than it is to decrease it. In short, you can move from a "Cape Canaveral" orbit to a "Baikonur" orbit, but the converse is very inefficient.

The inevitable price to pay, if you want to work with the Russians, was to build the ISS to orbit at 51°. The Russian launches would have had an easy life setting off from Baikonur, while the Shuttle would have to spend a little more power, but it could be done. Nevertheless, it was worth it, just to have the ISS programme get quickly under way. Although there was no real reason to try to beat the (former) (ex) Soviet rival, this political detail was swept under the carpet. In short, the fall of the Berlin Wall was responsible for moving the orbital inclination of the ISS—not directly, of course, but due to the difficulties NASA was having at that time, and the

Soviet space programme's reputation for excellence. All this was topped with a geopolitical "fig leaf" of a new collaboration between the superpowers.

The first piece of the ISS was the Russian module Zarya ("dawn" in Russian), which was launched from Baikonur in 1998. The plan, which was as always optimistic, was to have completed the assembly in orbit by 2003. Unfortunately, the reality was quite different: the ISS was eventually completed, at least in its basic parts, in 2011, more than eight years behind schedule. The programme cannot be called a success in terms of its being on time.

In the end, from 1998 to 2011, the building of the ISS took almost 200 trips, split between manned and unmanned flights. The latter, comprising more than 40 flights to the ISS, was mainly carried out by the Russians, using the *Progress* cargo vehicle; manned flights were to be divided approximately equally between the Shuttle (which also carried heavy loads on more than 30 flights) and a manned vehicle of the Russian Soyuz type, which carried almost no cargo. In short, without the Russians we would not have the ISS, even if NASA paid the bill for some of the Russian launches, especially the most recent ones. It goes without saying that after the end of the Shuttle programme at the end of 2011, the cost of a ticket for the ISS (and, above all, for the return flight …) had greatly increased, to something in excess of US$63 million.

Many other countries have participated in the construction of the ISS, 16 in total, especially Italy and other countries of the European Space Agency (ESA). Italy has been at the forefront, in contributing both structural elements and scientific instruments to the ISS since the 1990s. For example, Thales Alenia Space of Turin produced a major proportion of all pressurized habitable modules of the ISS. As always, Italian engineers and technicians have done a perfect job.

ESA has also helped with other important parts of the ISS—in particular, the Columbus laboratory, which is a cylinder almost 7 mt. long and 4.5 mt. in diameter. This is currently attached to the ISS where it forms the living quarters and working area of ESA astronauts. Since 2008, perhaps an even more important role arose for ESA: that of sending to the ISS a series of Automated Transfer Vehicles, (ATVs). These are unmanned, medium capacity vehicles capable of carrying eight tons of cargo to the ISS; they burn up in the atmosphere on their return to Earth. Although in the global economy of the ISS ATVs counts for little, Europe's involvement in its construction has justified it having a launch site in Kourou (French Guiana) and for the first time the ISS has been reached from a spaceport other than Baikonur or Cape Canaveral.

Also JAXA, the Japanese Space Agency, which has the bulk of material in orbit on the ISS Kibo laboratory (also taken up by the Shuttle), started operating its own independent transport vehicle, Kounotori 3, launched from Tanegashima (30°N), but it has made too few launches to have made a significant contribution to the work and survival of the ISS.

We are very interested, however, in the fact that the name chosen by Europe for the first ATV was quite rightly "Jules Verne"! "Did you hear this, Jules? What do you have to say about that?" And in honour of Jules Verne, they carried on board the cargo a copy of Hetzel's 1865 luxury edition of *"De la Terre à la Lune"* as well as an astronomical map, autographed by Jules Verne, showing the path that the astronauts would follow. And Jules, somewhere between flattered and embarrassed, says: "Ah, well, oui, oui, Hetzel … who was such a Shylock and did not want to make the deluxe edition for me—you know how publishers are … and as far as my astronautical calculations …—well, they were only a joke. But, anyway, thanks for remembering me to an audacious posterity, capable of such bold adventures!"

Of the four other ATVs so far (2015) launched by the ESA, one was named after Johannes Kepler, the great German astronomer of the seventeenth century, the father of the laws of planetary motion, the other after Edoardo Amaldi, the Italian physicist of the twentieth century, who was one of the fathers of ESA and the space research and activity in Italy, number 4 was named after Albert Einstein, perhaps the greatest scientist of all times, and the fifth after Gerges Lemaitre, the Belgian early twentieth century pastor who is credited with the original Big Bang idea. All names were chosen in recognition of the importance of collaboration and balance between pure and applied sciences in European countries.

But, having been painstakingly built, what purpose will really be served by having the ISS? How much can we really use it? And what, exactly, can it do for us? These are difficult questions. We have seen that the ISS was conceived with no strategic significance in mind, because the "evil empire" of the Soviet Union was no more there. Thus, it does not have any military significance, even any general, non-aggressive military purpose, and in any case the text of the construction agreement prohibits both these aspects from being in the nature of an international laboratory.

Does it have any scientific purpose? If not zero, then perhaps just a modest one in the study of microgravity and such like; but unfortunately, not enough to justify the enormous expense. From President Reagan onwards, US (and not only …) politicians had sought to make us believe in a scientific justification when trying to get the right reaction from the worldwide scientific community. At the very least "not with my money", as it was said then, and as it is still being correctly said today. It needs to be made clear that a proper investment into the scientific exploitation of the ISS needs to be done, given the *fait accompli* of its existence. But it must also be clear that spending on the ISS cannot be justified by science alone, and even less can its cost be charged to a science budget.

The ISS, instead, serves to a great extent, mankind's exploration of Sphere +1, and because it is a new habitat, it has been and continues to be a great training for human spaceflight. Since its completion, over two hundred astronauts have visited it from many different countries. It is, however, a low-altitude human space flight, which, while being above the atmosphere and in a zero gravity condition, remains within the Earth's magnetic field. It is precisely this magnetic field that blocks the entrance of cosmic radiation, which is so dangerous for astronauts and spacecraft going further from the Earth. The difference is important, because the real difficulty of human interplanetary flight, especially of the long-term one, will be the exposure to cosmic ionizing radiation from both the Sun and from outer space.

Above all, the ISS as served to show that together we could build it. That is, it stimulated the USA, Russia, Europe, Japan, Canada, Brazil and many other individual countries to develop a lot of innovative technology, while providing a new means of promoting industrial and global collaboration. Moreover, it has been a great financial investment. It has cost the taxpayer (especially in the USA) about $200 billion, which is, it should be remembered, half of what was spent by Obama in 2008 to save the banks from those who speculated on the mortgage of their houses. Italy, for example, has invested more than one billion Euros in the ISS: not a lot, but certainly a significant percentage of the budget of the Italian Space Agency (ASI), which was only formed in 1988.

Thanks to these investments, the ISS has produced "wealth" in the form of spin-offs, new jobs, economic and industrial development and so on, worth at least three times as much. This has been demonstrated by credible independent studies and is a point that should be considered for the future.

As for the future exploration of Sphere +1, in particular, we should now be somewhat concerned about it. And this is where things get complicated, because ideas concerning a global space community, with NASA at its head, though few, are confused. It is a fact that the ISS was completed only recently, but, though new-born, it is already dying. We have seen, for example, that NASA had to close the Shuttle programme. Since then, the only way in which astronauts can reach the station is Soviet-style, in the dear old, uncomfortable Soyuz, which is still in production and still working perfectly, thank goodness. Western astronauts who have used it (e.g. the Italians Roberto Vittori and Paolo Nespoli) have a rather mixed memory of it—excellent performance, but comfort on board seems to leave a little to be desired. Even in the Obama administration, the clear impression is that of a US progressive disengagement from the future of the ISS. Probably, in less than ten years from now, it will be expensively de-orbited.

NASA is now pretending to think of the "private sector" for the management of the ISS. It is an obvious bluff, at least for now. These private companies operate strictly under contract to, and under the control of, NASA, and are hardly different from the usual American aerospace giants, accustomed to fat and little-discussed government contracts (like all the other major aerospace companies around the world). In the case of the firm Space-X, for example, which produces the Falcon-9, a new vector for orbital transport, but which is not yet licensed for manned flight, NASA has provided a $1.6 billion contract for 12 flights up to 2015, with an option for more flights to get up to $3.1 billion by the end of the ISS's lifetime.

The company in question, however, while still conducting the first test flights, has already come up with an interesting source of income—so business goes. On the road to the ISS, it proposes to put into orbit hundreds or thousands of small, disposable urns containing the ashes of deceased persons whose relatives really want to see them in the sky … at the modest cost of a few thousand dollars per container (and the rocket carries a few thousand of them). Do not worry; however, they are not going to increase the number of orbiting objects. The containers will be "delivered" into an unstable orbit, so that they will return to Earth by burning up and leaving a poetic trail in the upper atmosphere. A kind of second cremation in space, with a shooting star included.

The Chinese, meanwhile, with determination and independence are also making Sphere +1 their business. After the flight of several "tachionauts", as the Chinese call their astronauts, their first objective has been to construct an orbiting base that, at least from the drawings that have arrived here, seems more akin to Mir than to the ISS. We have before us a major world power that is launching itself into Sphere +1, hopefully with new ideas, which are very much needed by the whole space community.

To pull out a few statistics on the status of the exploration of Sphere +1, let us start by saying that, as you read these words, there are certainly at least three people, but probably more, up to six, in circumterrestrial orbit. A little more than half a century after Gagarin's flight, the total number of people that have been in Earth orbit is in the order of 500, nearly half of them onto the Station, and these have been of many nationalities and both sexes (but always strongly male-dominated).

It is interesting at this point to make a comparison with Sphere 0. In 2003, fifty years after the conquest of Mt. Everest by Edmund Hillary and Tenzing Norgay, the number of people that reached the highest point on Earth was more or less the same, about 500. Again, these have been of many different nationalities, and again with a strong predominance of males (but for different reasons). Is going into space as difficult as going up Everest? Not from an athletic standpoint, but it does require some physical ability and determination that brings together the top players in the big business of exploration. However, if we take a cold look at the loss of life, we must say that going into space is much, much safer than climbing to the summit of Everest.

Today, more than half a century after man's conquest of Sphere +1, which is now permanently inhabited, what is there left to be explored? In fact, there is still quite a lot.

To start with, it remains for us to study in detail the physical environment that envelops our planet and which occupies a thin shell less than a tenth of the Earth's radius. Above all, we should study aspects of the potentially harmful radiation there, but we also need to work out how to continue our exploration of Sphere +1 without causing any further damage. In his early exploration, *Homo sapiens* committed serious errors, some of which are incurable. In those early years, but also later, very little attention was paid to the problem of space debris in orbit, some of it a perennial presence and potentially dangerous for astronauts in orbit, given the high speeds involved.

We will first have to catalogue all the space "junk" currently in orbit, in order to work out how to avoid it; we will then have to make sure that we do not produce any more of it; and perhaps, we should also devise some way of cleaning up some of the most affected orbits. It is a strange kind of exploration (similar again to what is now done in the Himalayas, i.e. removing the rubbish left behind by the old expeditions) but it is a necessary task in order to ensure a future for Sphere +1.

We also have yet to understand something perhaps even more important: how to continue on the road of international collaboration that began with the ISS, and which came about a little by chance and a little by the force of historical events. There is no doubt that if we continue on the road of the old and glorious COSPAR spirit, we might attain what today seems like a utopian dream and the only real direction in which we seem to lack progress: we need a World Space Agency for civilian space programmes. Coming from the whole Earth, it would speak for all of us in our quest of completing the exploration of Sphere +1 and safeguarding it for the future.

Chapter 6
Sphere +2 (or the Moon)

Exploration of the Moon

The next logical step in the exploration of space beyond the Earth as we go farther and farther away, after having conquered Sphere +1 once and for all, is to head for the Moon. Indeed, Jules Verne even skipped the intermediate stage and went directly to the Moon in 1865 with one of his best books, *From the Earth to the Moon*. The novel starts in Baltimore, in the USA, where in reality they must have been on their knees in that year, the one when the bloody and consuming civil war had ended. No matter: in a Florida resort, very close to Cape Canaveral, from where the Moon was actually reached a century later, a huge cannon was cast in bronze. It soon fired a special, gigantic shell to the Moon, complete with a crew aboard. But it will not land on the Moon: even Verne did not dare to imagine the intricacies of a Moon landing and subsequent lunar take-off. Rather, the shell-spaceship goes (balistically) into a lunar orbit, just as the two Apollo missions did in 1968–1969, and it will remain there until the end of the book. It takes another novel, *Around the Moon*, which appeared five years later, when it became clear that it was necessary to put things right. In the end, our heroes do return and, incidentally, land with an Ocean splashdown, just as the capsules in the Apollo project did.

Did von Braun and others learn from Jules? Who knows? Certainly they had read his work. In any case, in 1969, just eight years after Gagarin's first flight, Neil Armstrong, Buzz Aldrin landed on the moon, 380,000 km from the Earth's surface —that is a thousand times more distant than the orbit of Gagarin. And this is history. The story of the "conquest" of the Moon, or the exploration of Sphere +2, although very recent, has certain aspects that are likely to be forgotten, at least by the general public, but they are well worth remembering since they enhance even more the extraordinary undertaking it was and the incredible temporal and technological pace at which it happened.

In order to reach the Moon, we must first leave the Earth by finding the means to acquire the escape velocity of about 11 km/sec, which, although apparently similar to the 8 km/sec required to reach the circumterrestrial orbit of Sphere +1, is in fact much greater.

© Springer International Publishing Switzerland 2015
G.F. Bignami, *The Mystery of the Seven Spheres*,
DOI 10.1007/978-3-319-17004-6_6

Once again, the first objects which really left the grip of the Earth's gravity, that is who left Sphere 0, pierced through Sphere +1 and tried to get to Sphere +2 to explore the Moon, were made in the Soviet Union by Korolyov's team, with their powerful *Semiorka* that had been steadily improved. At the beginning of 1959, less than two years after Sputnik, the unmanned probe Luna 1 was launched. It was aimed at the Moon, but it missed the mark, becoming anyway the first man-made object to escape Earth's gravity. Another attempt was immediately made with Luna 2 on 12 September 1959 and that one worked well: it fell onto the surface of the Moon becoming the first man-made object to land on our natural satellite, or on any other extraterrestrial object. (It was followed, in the ensuing decades, by a lot of other "junk" landing there. Today, there are a total of more than 200 tons of man-made material, scattered everywhere on the Moon's face visible from Earth).

However, in 1959, the Moon was reached, while von Braun was again watching in anger, also because he now knew that his Jupiter rocket was no longer enough. Incredibly, just a few days later, there is another great Soviet achievement. On 5 October 1959, a new era was opened for humanity: this time Luna 3 accomplished a perfect trajectory from the Earth to the Moon, did not fall on the Moon but went around behind it, in what must be considered to be the first manoeuver in interplanetary navigation. It had on board a camera that worked and sent back to the Earth images of the far side of the Moon, that *Homo sapiens* (and all his predecessors and colleagues) had never seen and that, even today, we still do not see from Earth.

It turns out that the far side of the Moon, which we never see because of the synchronism between the rotation of the Moon itself and the revolution of the Moon around the Earth, is quite similar to the side of the Moon that always faces us, although it is somewhat more uneven. It would have been very strange if it were otherwise. However, on that date, 5 October 1959, the exploration of Sphere +2, and planetary exploration in general, began for *Homo sapiens*, who, just ten years later, would for the first time set his foot on a celestial body other than the Earth. And this time would be thanks to von Braun and NASA.

Meanwhile, in the late 1950s and early 1960s, the Cold War had not improved the situation between the two superpowers. Indeed, the Cuban crisis, the failed attempt of landing Cuban exiles at the Bay of Pigs (19 April 1961), sponsored by the CIA who were desperate to find a way to overthrow the Soviet-style communist regime of Fidel Castro, had brought the tension between the superpowers to very high levels. At that time, Nikita Khrushchev, the expert Party Secretary of Moscow, was trying to intimidate the young, and to him unknown, newly elected American President, John F. Kennedy.

But strangely enough, Kennedy was taken seriously on 25 May 1961. Just a few weeks after the crushing humiliation inflicted by Gagarin's spaceflight on 12 April, only partially remedied by Alan Shepard's short hop on 5 May, Kennedy announced: "*I believe that this nation should commit itself to achieving the goal, before this decade is out, of landing a man on the moon and returning him safely to the Earth. No single space project in this period will be more impressive to mankind, or more important for the long-range exploration of space; and none will be so difficult or expensive to accomplish"*. As a vision for the future, it had everything: to explore deep space, to impress the world (especially the Soviets) and to launch and accept an expensive and difficult challenge: these were the perfect ingredients in a speech to Congress. Except that, after the one miserable little suborbital flight of Shepard's a few days before, it seemed hard to believe that it would be possible to put a man on the Moon in less than ten years, and so there was panic at NASA. At this point, the US Space Agency had a duty to take up the challenge, but it clearly did not know how to do so, even if the promised budget were to be sums of money that had never before been seen. It was fortunate for NASA that it had access to von Braun and his team, in addition to a few bulldogs, who were hired to preside over and manage the work—first and foremost these was Rocco Petrone, the son of immigrant from Lucania a small region of the South of Italy. He became the director of the Apollo Moon-landing programme and was feared by everyone almost as much as von Braun was. He started the Apollo programme when everything still had to be invented. First, a serious plan of getting to and from the Moon; then a rocket of sufficient power; capsules for human transport; cruise engines for transit between the Earth and Moon; a vehicle for the descent and ascent to and from the lunar surface; spacesuits for the exploration of the lunar surface; antennae; computers; ... and much more.

The rest is well known. The Apollo programme is by far the best-documented exploration programme in the history of man, even without a Pigafetta with his magic pen on the ground or on board. Apollo followed two other NASA manned spaceflight programmes of the 1960s, in quick succession: the Mercury programme of 1961–1962, with suborbital flights and the first US orbital flight; and the Gemini programme of 1962–1966, with flights in Earth orbit testing the tandem capsule, manoeuvres and extravehicular activity. The Apollo programme, although formally started in 1961, in fact only really got under way a year later, in 1962.

The end result of the Apollo programme, which costed about US$120 billion in today's money, was that 12 people walked on the Moon, for an average of two days each; 381.7 kg of lunar rocks were brought back to Earth; and a total of 27 people were sent completely beyond the Earth's gravity. It is the best that mankind has managed to do so far in space exploration of the sphere of the Moon, Sphere +2 (or anywhere else).

Apollo was a programme of exploration, designed and managed in an exemplary manner, taking into account that everything, absolutely everything, had to be invented from scratch, and that difficulties surfaced immediately. For example, how does one go to the Moon? No one knew, of course. What was clear from the start was that, however big it might have been, a single rocket would not be enough to get from the Earth to the Moon, land, get out for a walk around, and then turn the engine around, re-light the motor and return to Earth.

Even when they had at their disposal, which luckily happened early in the programme, that incredible rocket—the Saturn V, the most potent object ever built by man—there was no hope of being able to opt for a scheme of a "direct" return trip with only a remotely "reasonable" chance of success (including the survival of the astronauts). The simple scheme was never even considered.

Von Braun, having the Saturn V in mind, knew that if he could build it, and even if it was still powered by just chemical propellant, it would have to do the journey step by step. At the beginning, he thought it was better to use a parking orbit around the Earth, where the various modules of the mission could be taken and assembled before leaving for the Moon. There was still the problem of descending towards the surface, landing and then taking off again, leaving everything behind except the module capable of returning itself to Earth, re-entering the atmosphere in a controlled manner and finally splashing down in the ocean.

Another NASA engineer, John Houbolt, hitherto unknown but who deserves to go down in history, proposed the revolutionary scheme of parking in a lunar orbit. After fierce debate, even von Braun was convinced, and the scheme was finally adopted in 1962 when the time left in which to implement it was running short. If von Braun had succeeded in winning everyone over to his original idea of an object in parked Earth orbit (instead of in lunar orbit), the subsequent history of the conquest of space would perhaps have been different. It is true that, in the case of the Apollo project, to get to the Moon, at least the first time, it would have taken two Saturn V rockets instead of one, and this turned out to be the decisive argument in favour of the lunar orbit.

But if a permanent module had been left orbiting the Earth as a starting point for all the Apollo missions, it would have been spot on the quasi-equatorial orbit to the Moon, which was easily accessible from Cape Canaveral. After the Apollo project, such a module would have remained there, as a kind of a precursor space station, small but operational 40 years before the current ISS, which was only completed in 2012. With a basic module, ready for the assembly of successive components and placed in the correct orbit, a mission to Mars would have been much more feasible, even in the short term. The fact remains that ifs and buts do not make history. Instead, we shall look at how the Apollo mission really happened, and in so doing, get a better general understanding of what it means just to get to and from the Moon.

The big Saturn V (110 mt. high, 10 mt. diameter) carried a set of two different double modules into Earth orbit: the Command and Service Module (CSM, two separate parts) and the Lunar Excursion Module (LEM), which in turn was composed of a descent module and a take-off module. At the start on the launch pad, the total weight of the mission was 3,090 tons, including a "payload" of 45 tons, to be carried to the Moon.

After take-off, while parked in Earth orbit, the lunar train is formed, consisting of the CSM and the LEM, and, while free of the atmosphere, the last tests are performed before leaving. Then, leaving the Earth's orbit, the train begins its traverse to the Moon, with an impulse from the last stage of the Saturn V, which is then jettisoned in space. Interestingly, the cruising speed chosen for the passage from Earth to Moon was just 1.5 km/sec. This was a relatively low speed, since the last stage of the Saturn V could have imparted a speed three times faster, allowing the target to be reached in a single day. But a spacecraft travelling that fast would have to pay a tremendous price on arrival at the Moon. The low gravity of the Moon and the absence of an atmosphere for aerodynamic braking would necessitate large motors (plus their fuel) to slow the spacecraft down sufficiently for it to attain a circumlunar orbit without crashing to the surface. In short, the choice to take three days, going at only 1.5 km/sec, was carefully planned and derived from the need to optimize the mass carried from the Earth to the Moon: taking a little more time required a much smaller mass of fuel.

Within three days, therefore, both, the CSM and the LEM, complete with all the fuel needed for insertion into lunar orbit, descent, ascent and return to Earth, totalling about 40 tons, reach the Moon. Using the Service Module engines and part of its fuel, the train brakes and inserts into a circular lunar orbit, at an altitude of about 100 km.

Once in lunar orbit, two astronauts prepare for landing on the Moon by moving from the CSM to the LEM through a special hatch. The LEM then detaches from the CSM, leaving in the third astronaut alone in lunar orbit to wait their return. With a manoeuvre, which is partly ballistic in order to save fuel, the LEM moves out of lunar orbit and, using its motors and all the fuel of the descent module, approaches the surface, on which it settles on its four legs.

At this point, the descent engine is off forever and will no longer be used. The two astronauts inside, don their appropriate spacesuits, depressurise the LEM with due caution, and make their way outside and down onto the surface of the Moon. Then, there are flags, pictures, television, strolls and the collection of samples—at least a gathering of whatever is at hand, without any scientific–geological plan, since any stone is a worthy sample, at least initially. The time spent in EVA (extravehicular activity), i.e. playing on the lunar surface, varied from 2 h 31 min for the first landing (Apollo 11) up to a total of more than 22 h for the last mission (Apollo 17), which also had on board a geologist–astronaut to collect a more rational selection of lunar samples.

After the EVA, the two astronauts return to the LEM, close the door properly, re-pressurise the cabin and remove their uncomfortable spacesuits. After a moment's rest, the long journey back begins. The first step, one of the many critical points of the mission, so-called single-point failures, is the ignition of the ascent module engine, without which they would be unable to leave the surface of the Moon. Fingers crossed as they wait to feel the kick in the pants that tells you that everything went well. The long-legged descent module is left on the Moon, with its engine off and empty, as the ascent module returns into lunar orbit, hopefully picking the right one, i.e. the same as the CSM, without which they would never get home.

Having found the CSM, there comes another critical point in the mission. They have to follow a perfect approach manoeuvre, firmly dock the LEM onto the CSM, open the hatch and rejoin the lone astronaut who has been waiting in lunar orbit. Finally, with all three again in the CSM, what remains of the LEM (i.e. the ascent module) is jettisoned and allowed to fall back onto the Moon. The CSM uses its engine and the fuel carried from the Earth to set them off again on the cruise phase of their return to Earth. Their mass as they leave lunar orbit is now about 25 tons.

After two or three days of cruising, always conserving fuel, the final approach is made towards the Earth. Shortly before the critical manoeuvre to re-enter the atmosphere, the Service Module with its engines, that are now no longer needed, is jettisoned and allowed to burn up in the Earth's atmosphere. The Command Module, with the three astronauts and the precious lunar rocks, with a total mass of about 5 tons, is all that remains of the mission, which, remember, departed from Earth with a mass of more than 3,000 tons.

But it is not over: the CM must enter the atmosphere at exactly the correct angle: too steep and it will burn up like a match; too shallow and it will bounce off the atmosphere like a skimming stone on water, and be lost forever towards the Sun. Not only that, but immediately before entering the atmosphere, the CM, which has a conical shape, has to turn around and present the base of the cone to the direction of motion. The base is where the heat shield is located, and the shield is the only thing capable of saving the astronauts from certain death as a shooting star.

Decelerating violently in the upper atmosphere, the CM becomes surrounded by a fiery cloud of gas that prevents any communication with mission control. Then comes the last critical point: the opening of the parachutes. Remember that the parachutes, carefully packed in their housing, have faithfully followed the CM to the Moon and back, undergoing accelerations, temperature changes, etc.—we hope that the journey has not been too detrimental to them. Of course, they have to open at the right time: too early and they would tear apart because of the high speed; too late and there would not be not enough time for them to deploy properly. Finally, the splashdown—a manoeuver that is much less sweet than you might think: "We fall into the water like a ton of bricks", said the members of several crews. At that point, lulled by the long Pacific waves, they await being picked up (while they hope that the CM does not sink). They are recovered from the sea by frogmen and a rubber boat and are transferred to a helicopter. All the while, they are dressed in uncomfortable airtight suits, to prevent contamination of our planet by hypothetical Moon germs, until they are quarantined in a container aboard the aircraft carrier on duty. There, finally, they are able to take a shower, drink a bourbon and eat a steak.

"Dear Jules," I ask, "What do you think of an actual mission to go to and return from the Moon? Complicated, eh?" to which Jules responds "Well, yes. The reality is impressive, but I too had planned the trip around the Moon with the return ending in a splashdown—but I had not dared to land my astronauts inside a cannon ball on lunar soil. Although Méliès did it in his beautiful film!" I must say he is quite right: his first book seems like a test run of the Apollo 8 mission, the first that took three men to the Moon, around it and back.

We wanted to relate the story of a typical Apollo mission and so compare fantasy with reality in the exploration of the unknown, especially of space. Although we are just trying to explore Sphere +2, and thus we are still very close to the Earth, there are several things that can go wrong. For example, when the first mission of the famous Apollo 11 reached its destination, nearly everything went wrong. During the final stages of landing, the on-board computer suffered a breakdown, even if it was a ridiculously simple computer by today's standards, with much less computing capacity and memory than a current conventional mobile phone. Consequently, Neil Armstrong had to pilot the descent manually, navigating between clouds of dust that obscured his view of the landing site. One must remember that a simple rock under one of the LEM's legs would have tilted the structure enough to prevent the return launch. The fact is that Neil positioned the LEM on the surface with a handful of seconds (some say ten; some say twenty) worth of fuel left for the engine to complete its descent. If they had run out of fuel before landing, they would have fallen like a stone (and would never have left the lunar surface). Then, while donning their spacesuits inside the LEM, without realizing it, they physically broke the switch that was to arm the ignition of the engine for lift-off. Later, at the right moment, Buzz Aldrin coolly managed to start the engine with the tip of a ballpoint pen. Finally, there was a tragicomic incident during splashdown when, for unknown reasons, the Command Module in the water tipped over. The crew, who had already been well buffeted by the return and who were prostrate from finding themselves again in the grip of a strong gravitational field after more than a week away, found themselves hanging upside down, strapped in their seats, watching the blue waters of the Pacific through the sub-merged portholes, hoping that they would stay afloat and that they would at least be spared seasickness.

More serious was the case of Apollo 13 a year later, made famous by a beautiful film. In that case, the explosion of an oxygen tank rendered the Service Module unusable during the outward bound cruise phase. Fortunately, the astronauts were able to use the LEM, which was uncomfortable but functional, and in this fashion managed to get back home, while having to re-calculate the orbital flight for the unexpected return and mentally thanking the engineer Houbolt for the LEM he had wanted and invented. Yes, on the whole, the Apollo programme was incredibly lucky, given the very limited safety margins adopted, and given the speed at which it was done, just to get there before the Russians. Today, no Space Agency, least of all NASA, would ever accept such high risks.

But today, almost half a century after the Apollo missions, and especially for the future, what remains to be explored in Sphere +2, i.e. the Moon and its sur-roundings? Certainly, the answer is "almost everything". Let us start with the lunar surface: we now have excellent cartography of the Moon, obtained thanks to some orbital missions at low and very low altitudes, down to just 20 km. Because there is no atmosphere on the Moon, satellites can fly low, as long as they can be sure of avoiding the mountains. But what we do not know, for example, is where the deposits of water are that should have been left by the ice of comets that have impacted on the Moon.

Of course, water can also be extracted from lunar rocks by a complicated and expensive chemical process (which also requires a lot of energy), but if it is found in the form of the millions of tons of ice that make up a comet, so much the better. And once enough water has been found, it is no longer science fiction to imagine at least medium-term survival on the Moon. Of course, this would be true as long as one also has power available, probably by a nuclear generator, compact enough and easy enough to assemble—an object which, however, remains to be invented.

Well, we are not talking of survival on the surface of the Moon, since it would not be possible to survive there more than a few weeks; nor on the surface of Mars, as we shall see. Those science fiction films that show elegant transparent "domes" used as bases will never come true (well, never say "never"…). But cosmic rays that come everywhere from the sky, including the Sun, are rapidly fatal, without a thick enough protection, especially during solar storms, to all forms of life unprotected by an atmosphere and a magnetic field like the Earth's. Also owing to possible dangerous meteorite hits, also unchecked by any atmosphere, the only possibility, at least initially, will be to settle down safely underground. A few metres of rock should be enough.

This consideration, unfortunately, makes it very unattractive, if not impossible, to create tourist resorts on the Moon, a place where, to stay healthy, life must be neither easy nor fun. Too bad, because apart from the fun of exploration, there were those who imagined tall, slender buildings as holiday hotels; structures conceivable only in the low-gravity environment of the Moon. While they could be beautiful, fantastic shapes, and the dream of every architect constrained by the Earth's gravity and the structural relations which it imposes on buildings, they would, alas, be exposed to radiation and, let us not forget, to meteors big and small, which would have no atmosphere in which to burn up before impact. It is a shame to have to abandon the idea, which has already been circulated by the organizers of extreme holidays, of pressurized structures with climbing walls where everyone could have easily climbed extremely difficult routes, or with springboards from which it would be easy to jump to enormous heights.

More realistically, it would be interesting to explore the lunar soil to find the lava tubes that apparently exist on the Moon, similar to those on Earth, for example around Mt. Etna in Sicily. If one were to be found that was spacious enough, it might be used for creating a sealed environment in which an air supply could be made to circulate, producing something close to the Earth's atmospheric pressure, humidity and temperature. In short, we could try to recreate a terrestrial environment; it would be difficult, but not impossible. Having thus found the right place for it, and figured out how to build it, a lunar base would not then be science fiction, but real, and capable of hosting a new species of *Homo sapiens selenicus*, although we must ask ourselves what we would do there. Certainly, we can discard the idea, proposed by the then *science advisor* to President G.W. Bush in 2008, of making it into a departure and arrival base for spaceships directed elsewhere, for example to Mars. It would be an act of gross stupidity, and we can see why. First, to land on the Moon, one has to spend quite a lot of energy, simply to avoid an otherwise inevitably catastrophic landing. In fact, in the absence of any atmosphere, parachutes do not work, and so to avoid crashing into the ground, one must use suitable brake engines or reverse thrusters, such as those on the Apollo LEM, for example. For an interplanetary cruise ship, the braking system and its fuel would constitute a significant mass, which would have to be brought from Earth, so reducing the payload of the ship, etc.

If that were not enough, to leave the Moon one must still spend yet more energy. Although the gravitational field of the Moon is smaller than the Earth's, it is not zero, and it has to be overcome before going elsewhere. The amount of resources needed to land and re-launch, the necessary fuel and its rockets, motors, nozzles, etc. (the little LEM weighed only 16 tons), constitutes such a useless overload for any clipper on the Earth–Mars route, which would render the whole operation unfeasible.

No, a lunar base cannot serve as an intermediate stop for interplanetary routes, but one could at least imagine a massive production of fuel on the Moon, which, at least for now, could in principle be supported by a lunar base in a lava tube.

So what are the prospects for *Homo selenicus*? Unfortunately, the chances of profitably exploiting a lunar mine are ridiculously small, given the transport costs. And then, that the Moon itself is a piece of the Earth, as is now well established. To sink a mine on the Moon would be, more or less, as good as sinking it in any random place on Earth: it would be difficult to expect spectacular results.

The same applies to precious metals, which are present on the Moon but certainly not in sufficient quantities to encourage any lunar gold rushes, which would be much more expensive and dangerous than those of Alaska or California at the end of the nineteenth century. What about diamonds? Who knows? There has certainly been some volcanic activity on the Moon—perhaps an exploration of volcanic chimneys could be interesting. If there were diamonds in the lunar core, Verne would have immediately said that they would be enormous because the lower gravity would have allowed them to grow bigger ... fantasies. Back to reality —mining the Moon is not an activity in which to invest.

Another thing that has been discussed is the famous helium-3 (He-3), a helium isotope that is rare on Earth but which would be very useful as a fuel for nuclear fusion, if we knew how to do it. Some think that helium-3 should be particularly abundant on the surface of the Moon, as a result of the bombardment of solar cosmic rays. It is unknown if this is true, but even if it were, its collection would not be easy. And we do not want to encourage surface extraction on the Moon, or strip mining, as the Americans call it. Such activity on Earth is definitely frowned upon, if not prohibited, and we will not begin to deface a celestial body as soon as we have arrived.

However, we will not know how to produce energy from fusion for many years, and thus, any lack of helium-3 as a fuel is not the problem. Normal hydrogen isotopes are abundant enough in water, which itself is abundant enough on Earth. To collect helium-3 from lunar sources and then bring it back to Earth to create energy is a hoax that serves only to confuse. Our earthly means should be more than adequate to the task of obtaining fusion, at long last.

What, however, surely could and must be done on the surface of the Moon is scientific research. Not only to better understand our satellite, which is in itself an important goal, but above all, to use the Moon as a base for an astronomical observatory of the Universe. The lack of any atmosphere, the total absence of stray light and the directional stability (much greater than that of any satellite or observatory) are some of the characteristics that make it very interesting to use the Moon as a base for scientific observations. Of course, the large telescopes would have to be carried to the Moon and then installed and managed. It does not seem, at least for the near future, an easy job, while living in a lava tube.

More attractive and feasible seems to be an observatory that would make use of radio waves, a radio telescope, placed on the far side of the Moon. A radio telescope is easier to mount than an optical telescope and perhaps is something that could be done robotically. And it would have the huge advantage that, by looking at the

Universe from the far side of the Moon, it would be completely shielded from the continuous and growing radio noise emitted from the Earth. With radio noise from military radars, large broadcasters, and the private radio around the corner, it has become increasingly difficult for radio astronomers to find a frequency in which to look at the cosmos undisturbed. The side of the Moon in the electromagnetic shadow of Earth would be an ideal site from which to explore the deep cosmos. Perhaps, it might even be possible to detect those signals sent by intelligent ETs, who must exist somewhere and who could be trying to make themselves heard.

Then, there is the exploration of the Moon's surroundings, which are an integral part of Sphere +2. Not far from the Moon, there is a point that is gravitationally very interesting, a point to which Verne also refers. This is the point, or rather points, technically called "Lagrange" or "libration" points. These are the points at which a material object is "free" because it is under the equal gravitational pulls of the Earth and the Moon, and so it is poised not knowing whether to fall on one or the other celestial body. It is obvious that such a point exists, going from the Earth to the Moon, but the mathematical analysis necessary to find the libration points is quite detailed and difficult. The first solution was found around 1760 by the Italian French mathematician Jean-Louis Lagrange, who demonstrated that there are three points in question, two in the vicinity of the Moon and one near the Earth.

"*Du reste, Jules,*" I say to my inseparable friend "even you used the concept of the balance between the two gravitational fields in '*Around the Moon*' ..." "*Eh oui,* and to think that I knew nothing of Lagrange and its formalization, which also dates back almost a hundred years before my novel ... I just called it *le point neutre,* I thought that would give the idea ..." "*Chapeau*" I say "hats off ..."; however, of particular interest to us are the two libration points closest to the Moon, which are about 50,000 km above the lunar surface, one on the side of the Earth, and the other symmetrically opposite on the other side. They are interesting for their essential lack of gravity. There are those who have argued that these points would accumulate a certain amount of space junk, comprising various meteors that, once there, would not know where to go. Only imagination, of course: to leave these points only requires a small velocity, such as a normal meteor certainly has. In particular, the libration point in the vicinity of the nearside of the Moon might be the natural home of a spaceport for interplanetary (or interstellar) flights departing from Earth. As we will see in more detail for the exploration of Sphere +3, a nuclear-powered spaceship for travel into deep space, such as going to Mars, requires a long period for its assembly, after being transported in pieces from Earth.

The ideal thing would be to start an interplanetary trip from a point in space where it is not necessary to expend much energy to get underway, that is precisely the definition of a point of libration. Other than a base on the lunar surface, or from a parking orbit of the Earth or Moon, all situations call for an unnecessary consumption of energy: the point of libration between the Moon and the Earth, a little more than two days' journey from the launch site on Earth, is the ideal place to establish a spaceship yard, a space port.

We can be fairly certain that the point of libration, and its as yet unexplored potential, is currently being studied by NASA (and perhaps others). Starting from the Earth, it should not take long to build a floating construction site in order to go around the solar system and, for the first time, take *Homo selenicus* beyond the Moon.

But in expressing this certainty, which is more of a hope, we must ask ourselves why, since 1972, no *Homo sapiens* have visited the Moon or its surroundings, or more generally, why no human being, after the 27 who went on the 9 extraterrestrial flights of Apollo, has felt on their body the acceleration that takes one to the fateful speed of 11 km/sec, into a trajectory that does not return to Earth.

Certainly, there is no longer a Soviet empire and, in today's world, for the USA, the Chinese are not yet formidable competitors in the space race. The Chinese have yet to go to the Moon, and if they ever succeed (in something which is not easy, as we have seen), they will not only be second, they will have arrived at least half a century later.

The answer to the question "Why have we not returned to the Moon, or gone elsewhere?" is disconcertingly simple: we have not so far managed to combine our strengths and resources, both practical and cultural, into a common project at a global level that will give, in equal measure, benefits to all *sapiens* on Earth. But it can be done.

Part II
Mars, Planets, Stars

Prologue 2

Adieu Jules, faithful companion of all my adventures on land, sea and sky, up to the Moon. We shall now be going to other Spheres, much, much farther away, Spheres you never even dared to dream for. Today, nearly one and a half centuries after your beautiful imagination works, it is only fair for present and future exploration to carry us, in our minds at least, deeper into the sky. It is something that is required from us by our own history, continuously and with an ever accelerating pace since we came out of Africa, children of one single African Eve, who lived 143,000 years ago.

With tens of robotic space missions, we have already been beyond the Moon and more, actually throughout the whole solar system. Actually, four man-made objects, four NASA interplanetary probes, have already flown out of the Solar System and are now exploring interstellar space.

We shall have a tale to tell of the robotic exploration of Sphere +3, a Sphere to which we shall endow a considerable "thickness": it will extend from the orbit of Mars to the border of the Solar System.

The inner boundary of Sphere +3 is the orbit of Mars which is a thousand times more distant than Sphere +2, that is to say the orbit of the Moon. This, we remember, is a thousand times greater than the distance of Sphere +1 from the surface of the Earth. Moreover, Sphere +3 will have its external boundary at the border of the Solar System, a few thousand times further away than the orbit of Mars. There is an agreeable music of symmetry in the dimensions of our Spheres.

After robotic crafts, which have been travelling around in the Solar System for some time now, it will be up to us to explore Sphere +3. Better be prepared which means, in our case, to be able to describe, or to imagine, how this could be done. Few people know that von Braun, just after the Moon had been reached using chemical propulsion, in 1969 was ready to organise a human expedition to Mars. It was a 12–15-year plan and would have used what at the time was still nearly science fiction: a fleet of nuclear-propelled space ships using nuclear fission engine. Von Braun only knew nuclear energy as it was known in the 1960s and in general terms, without great details. Yet he knew about it, and he was the only one with the charisma and the courage of proposing it in a credible way, using a realistic and well-developed plan. So, for Sphere +3, we shall trust him: "zu befehl, Sturmbannfuhrer von Braun".

For the human exploration of Mars, however, it would be easy to see how today it will not be necessary to go beyond the current limit of science, i.e., our current knowledge of physics and technology of fission nuclear propulsion should be enough. In order to make it all happen all that would be necessary would be a sentence like that uttered by Kennedy in 1961, naturally delivered with the same authority. In other words, all that is needed is that we really want it, and in 20 years, or even less, we would be on Mars.

A totally different issue is that of bringing a human crew beyond Mars and back, in a reasonable amount of time. We shall see that we could talk about this too without a need to invoke science fiction. We shall, maybe, push a little beyond the current limit of science, but we shall do so following a rigorous physical criterion regarding, for example, the propulsion energetic efficiency. Beyond Mars (and maybe Jupiter), we shall invoke well-known type of physics, that of nuclear fusion, even if the relevant necessary technology is still totally beyond our reach. Wernher here will not be able to help us for what concerns physics. But he will remain the perfect leader (Fuhrer) for the planning and political management of the organisation and financing of an interplanetary mission, not to mention the strenuous crew training that will be necessary.

Further out, when we shall exit the Solar System through the outer edge of Sphere +3, we shall feel somewhat lonely, without Jules Verne nor Wernher von Braun. But real explorers are not afraid of the unknown, on the contrary they look for it, however, hard it may look. The nearest stars are only a handful of light years away from us. Nothing on a universal scale, but at the same time an enormous amount compared to our propulsion and transport capabilities onto the timescale of our lives. Leaving Sphere +3, we shall arbitrarily define the radius of Sphere +4, to one of the fixed stars, as 20 light years, we shall later see why. We shall also look at the means and the motivations to carry human exploration so far away. It is obvious that for Sphere +4 we do not yet have the necessary technology. And even for the necessary physics, we can only invoke a few fundamental laws, at least those which we know best. For the rest, we shall try and go beyond the limits of science and technology, as they are currently known, realistically trying for an imaginative extrapolation.

If you think about it, we feel much more justified in doing so than Ciolkovskij, who did it at the end of the nineteenth century, well before any human being had left Earth. The same applies to von Braun, born over a century ago. Ciolkovskij was the man who said that the Earth is the cradle of men, but one, sooner or later, must always get out of the cradle. Von Braun was the man who gave us the Moon and could have given us Mars had he only had a chance. Both of them, had they mastered today's science and technology, would not hesitate.

Us, mere mortals, are in need of a higher help and inspiration. Luckily we shall find both right there where the going gets hard at the extreme boundary of the Solar

System. For now, let us imagine we shall be able to use, in a not too distant day, the most efficient energy form yet known to us in nature: matter–antimatter annihilation.

Let us anticipate here that, for reasons not fully understood by modern physics, when a particle of matter, for example, the nucleus of a hydrogen atom meets its antiparticle, an antiproton, the masses of both particles "annihilate". This means that, according to Einstein's equation $E = mc^2$, both masses turn into energy, multiplied by the square of the velocity of light, a rather hefty number.... . Among the many problems to be tackled in order to exploit such an enormous energy, the most important is that in the annihilation process energy is produced mostly in the form of gamma rays. Admittedly, these are more complicated than petrol to be used as propulsion; but, no matter, we shall find a way of doing it.

However, we are talking about an energy source capable of producing an energy per unit fuel mass about a thousand times bigger than nuclear fusion. Of course, none, so far, has managed to exploit matter–antimatter annihilation energy. We do not know if we will manage it, especially for propelling a spaceship. But, leaving science fiction aside, and looking at the problem from the outside, we observe that annihilation is the last step known to us in an increasing sequence of energy production processes. It is a sequence which translates into our capacity of pushing to higher and higher speed a ship fit to carry future *Homo sapiens* towards its ultimate exploration goals.

In orbit around the Earth and then on the Moon, we went exploiting what best chemistry could offer us. We are talking about solid and liquid chemical propellants but above all of the grandiose water molecule and its incredible properties. The needed velocities and masses for ships allowing us to survive from here to the Moon have been within our reach for many decades now. To go to Mars, and maybe a little bit beyond, we know that speed must be up by an order of magnitude and the same for ships' masses. However, the fission of heavy nuclei, a process we master very well, could bring us to Mars and beyond starting tomorrow, if we only decide to do it.

To explore the whole Solar System in times reasonably compatible with human lives, we should take the next step, both in velocity and thrust. Nuclear fusion would allow it if only we knew how to exploit and harness it, in particular for propelling a ship of adequate dimension. We shall see that a few ideas are already being thrown around. After fusion, the next ladder in the energy scale could come from controlled matter–antimatter annihilation. It could bring our crazy explorers to our nearest stars but not beyond. It would open for us Sphere +4, the one which could give us a new Earth on the timescale of a few generations. Nothing, if you compared it to how long it took for the children of the African Eve to reach a new continent.

However, to reach Sphere +4 we shall need some inspiration and leadership. Just like we left Jules at the Moon Sphere, we shall be leaving Wernher beyond Mars and the inner Solar System. Someone special is waiting for us to lead us to the most difficult jump towards the stars.

Chapter 7
Sphere +3 Inner (or of the Mars)

Homo sapiens on Mars

Human exploration in space is at a standstill since 1972, from the moment when we reached the Sphere +2, the Moon. Seen from the perspective of the unstoppable expansion of *Homo sapiens* over the last 70,000 years, this forty-year hiatus, which seems to us long, is actually negligible. There is no doubt that very soon we will leave Sphere +2, arrive on Mars and then continue on much farther. It would be too strange to think that today, at this precise moment in our history, we, the "cosmopolitan invasive" human, might bring our drive for exploration to a halt. To think that one lives in a "special" age is always dangerous and generally unhistorical. And to think that one lives in a unique, unrepeatable moment in the long history of mankind seems really simplistic, or else conceited, because it is unscientific to imagine one's own time to be special simply because one finds oneself in it. It is a little like thinking that the Earth is the centre of universe; or that the village where you were born is the most beautiful of all.

No, we prefer to think that we are not living in an exceptional moment in human history, also because we would not know how to explain such a fact. No other characteristic of *sapiens* has changed, either for better or for worse. On the one hand, he/she continues to be inventive and imaginative, as well as methodical, tenacious and courageous: the continuous acceleration of scientific and cultural progress is there to prove it. And the mindset behind this acceleration is very similar to the one that has always driven him on and on to explore more and more. On the other hand, *Homo sapiens* continues to maintain those characteristics that we now consider largely "negative", such as greed for gain, aggression, religious fanaticism, and thirst for power. Apart from today's moral considerations, these explorative characteristics have been important factors that lay behind the success of our species, precisely because they are fundamentally explorative, and they have taken us as far as the sphere of the Moon. Globally, we have not changed, and so let our incurable explorers do what they must. And this is why we will go to Mars: what is left to us is just to decide how and when to go.

© Springer International Publishing Switzerland 2015
G.F. Bignami, *The Mystery of the Seven Spheres*,
DOI 10.1007/978-3-319-17004-6_7

The first big opportunity was lost during that happy August of 1969, a few days after the landing of the first man on the Moon. Few people know that in those days of glory, von Braun, the hero of the moment, appeared before the US Space Commission, which had the power to approve new space programmes. Riding on the wave of the world's enthusiastic reception of America's conquest of the Moon, von Braun made a very short and punchy speech, illustrated with some beautiful slides that are still available in the historical archives of NASA. He explained how, after the Moon, the USA could be the first to land on Mars by 1981.

Von Braun showed that to get to and from Mars it would be necessary and sufficient to use a fleet of dozens of "nuclear-powered" spaceships, an expedition which could be done at the cost of a "Minor operation in any theatre of war" (which is true even today). Von Braun had even thought of a system for landing, involving approaching the Martian polar cap at a low angle and landing on the ice with special skids deployed at the right moment—like on ski slopes. The project was rejected, apparently by just a few votes, but not because of the ski-based landing gear: simply because the USA was at the height of its devastating war in Vietnam and funding for space programmes had to be cut, as happened just a few years later to the Apollo programme itself. Had that day in August of 1969 returned a "Yes" vote, we would already have been to Mars, maybe not in 1981, but certainly by the end of the second millennium.

We have spoken earlier of von Braun in the days of Sputnik and the Semiorka and of the race to the Moon. But it is interesting also to review who this genius of the missile was and where he came from, as he had a very difficult past. Wernher von Braun's father, Baron Magnus von Braun, once said that what he remembered most about Wernher as a small boy was that it had been absolutely impossible to educate him according to his own strict principles. He gave up, resigning himself to pay the bills for the broken windows and the flower beds destroyed by the young Wernher's experiments with rudimentary rockets. Von Braun had been born in 1912, in a village in the east of Silesia, which became Poland at the end of World War I. The family moved to Berlin, where the young Wernher continued on the same road: the good Berliners were shaken by explosions coming from his rocket-powered car with which, at the age of 12, he and his companions were trying to break a speed record on a Berlin street. In the end, he was arrested and had to be rescued by his father, the Baron. But despite these exploits, he also made music with much finesse and played well both the piano and the cello throughout his life.

The real breakthrough came when, after Catholic confirmation at the age of thirteen, his mother, the noblewoman Emmy von Quistorp, gave him a telescope. In general, he very much took after his mother, who was very intelligent. When he was ten years old, she asked him what he wanted to be as a grown up. Wernher replied: "I want to work with others to help turn the wheel of progress". In 1927, came the second high point in his young life: he read the book *Die Rakete zu den Planetenräumen* (Rockets towards Interplanetary Spaces) by Hermann Oberth, the pioneer of space propulsion, both in his native Germany and at a world level. As soon as he finished reading it, he wrote, at the age of 15, to the great scientist saying that he too (!) believed in space propulsion and had written an essay about it. Oberth replied, encouraging him to continue.

He studied engineering in Berlin and received his doctorate very early, in 1934, with a thesis on propulsion, a large part of which was classified by the armed forces and was only made public in 1960. The fact is that Wernher had begun to work on the development of rockets with possible military applications, since the development of civil rockets was banned by the regime. The evidence of von Braun's links with the Nazi party was, rather understandably, swept under the carpet during his years with the Americans. We do not know whether he joined the party in 1932 or 1937. What is certain is that von Braun was a fervent Nazi by choice and belief, not only out of necessity, as he tried to say later on several occasions. He certainly became an SS officer at the beginning of World War II and ended his career in 1945 with the rank of SS *Sturmbannführer* (Major), with numerous important decorations, including the *Kriegsverdienstkreuz* (War Merit Cross) and *Ritterkreuz* (Knight's Cross, first class, with swords). Not ordinary decorations indeed, since they were delivered personally by Hitler, who had also conferred him the title of "*Herr Professor*". His decorations, however, were well deserved from the point of view of the regime: in 1937 (at 25!), he was already the technical director of the Centre for Military Missiles on Peenemünde, an island in the Baltic which he chose, it seems, because his grandfather used to go there duck shooting. From there, the V1 flying bombs were launched towards England and Belgium; and from there, in September 1944, the first rocket-powered *Vergeltungswaffe-2* were launched, the second deadly "weapon of revenge" of the doomed Reich. To build the V2, under Allied bombing, von Braun did not hesitate to use, quite ruthlessly, slave workers including Jews, prisoners of war, and other so-called enemies of the regime, that infernal system of production that the punitive Reich put at his disposal. He seems to have approved the system, though, since he assisted in administering cruel corporal punishments, including hangings by chains, and ordering executions to set an example. All this was later washed over by the sponge of the victorious Americans; but those who suffered the cruelties did not forget, such as the communist French partisan Guy Morand, who remembered well, when he testified to the fact that von Braun kept ordering the executioner to whip him harder.

At the end of the war, it was very clear to SS Major von Braun to whom it was preferable to surrender. With a train full of V2 parts, they headed south towards the American zone, carefully avoiding the Russians, who were not renowned for their tenderness towards the SS. The Americans immediately understood the potential interest of their prisoner, and von Braun, his team and all the V2 material and equipment left for the USA, ending up in Texas. There, the German engineers immediately began work, to demonstrate to their new masters, the US Army, how good were the Germans of Peenemunde, even though higher authorities had ordered them to be kept in isolation. They considered themselves PoPs (Prisoners of Peace), even though they had written on their backs PoW (Prisoner of War), but they soon removed that rather infamous uniform, which was so evocative of a very difficult past. However, at the end of 1946, just over two years after the launch of the first V2 at Peenemunde, the first V2 was launched from US soil. As a reward, it seems, von Braun was granted a very special permission to return to Germany, where, on 1 March 1947, at the age of 35 he married his mother's cousin, the 18-year-old Maria

Luise von Quistorp. They had two daughters and a son, all still alive at the time of writing. In the years from 1946 to 1949, von Braun began his career as a writer and popularizer of science and produced, among other things, the original German version of his *"Mars Project"*, all the while continuing with his activity in missile research. In 1950, while the Korean War was still going on, he was transferred to the Redstone Arsenal, Huntsville, Alabama, where he remained for twenty years and where in 1955 he became a US citizen (thanks to the cleansing action of the sponge!). In the early 1950s, he is seen in *"Collier's Weekly"* and *"The New Yorker"* magazine, and on television with Walt Disney. But it was really at Huntsville where von Braun had his first success: the Redstone rocket, the first ballistic missile, which, in the late 1950s, was successfully tested as the main delivery system for atomic bombs.

Returning to Mars, von Braun's draft report, soon to become a book in English, on how to reach the red planet was credible and detailed and is still a very enjoyable read. But in one critical point, it falls short: the fleet of spacecraft planned for Mars is propelled, says von Braun, by nuclear-powered engines, but he does not explain how such engines could operate in a spaceship. Was this because of security reasons during the Cold War? Perhaps; or more likely, von Braun, despite having realized that nuclear power was needed, did not know exactly how to make it safe enough, and powerful and compact enough to install it on a manned spaceship. Today, after several decades of physics and nuclear engineering, we know how to construct a similar engine, even if we have not done so yet. And we are now so incredibly much further ahead of the Apollo project, in terms of computing capacity and miniaturization, data transmission, mission analysis, interplanetary navigation, servo systems and mechanisms for landing, life support systems of astronauts and much more.

So technically, we are now much closer to Mars than Kennedy was to the Moon in 1961, when he announced that "this nation … before this decade is out… would go to and return safely from the Moon". He just had no idea how it was to be done; and neither, of course, did NASA. But once it had been decided, in order to obey their president, NASA realized that they had to neglect a lot of the normal safety criteria for human flight. And, if we are honest, it was just a case of sheer good luck if Apollo escaped losing at least one crew in flight. It was a tragic misfortune that three astronauts did lose their lives in a ground test of the Apollo 1 capsule, which burned on the ground when, in an atmosphere of pure oxygen, a spark was enough to trigger a devastating fire. We should always remember, however, that we (meaning *Homo sapiens*) landed on the Moon because Kennedy had issued the challenge *"before this decade is out"*: had he said *"two decades"* we would never have gone there. The Apollo programme was in fact progressively stopped, starting in 1969, under the Nixon administration, but the work was already too far advanced for it to be unsuccessful. Had it been stopped earlier, there would have been no Moon, only the Vietnam War. In short, not enough time had been given to the project for it to be cancelled. A lesson to remember.

Certainly, in terms of exploration, going to Mars today would be much less risky than it was to circumnavigate the world five centuries ago for Magellan (who of course died on the voyage, together with most of his ship's company), or a century ago by Amundsen and Scott who reached the South Pole on foot (and here too many lives were lost), not to mention countless other expeditions. Although the dangers of exploration have never ceased, a mission to Mars can now be made much safer than the Apollo Project has been.

So, let us see how a manned mission to Mars really could be done. The situation differs from the Apollo missions because the distance from Earth to Mars is about a thousand times that from the Earth to the Moon. We know that we could never do such a journey in a week; but to begin with, let us set ourselves the limit of staying in deep space for one year, which would cause reasonably little harm. The necessary speeds are immediately calculable as being in the order of tens of km/s, depending on the relative positions of the two planets, and this is 10–20 times faster than Apollo. For a crew to survive such a trip through the Solar System for a year, they would need a ship with anti-radiation shielding, and large enough to include living space with a kitchen, a gym, maybe a small hydroponic vegetable garden, etc. All in all, the ship should be something of the size of a large Airbus, which means many tens of times the mass of the Apollo Command module. The inevitable result is that the combination of distance, maximum allowed transit time and global mass requires, for Mars, an energy source much more "efficient" than what can be achieved by the chemistry of combustion, i.e. an energy source with a much higher yield per unit mass.

In physical terms, this means switching from chemical to nuclear power, as von Braun well knew. The fission of uranium-235, for example, delivers 8×10^7 MJ/kg, which is almost a million times more energy efficient than the best performing chemical process—i.e. the recombination of the water molecule. In technical terms, nuclear power means being able to combine in a single engine both "thrust" and "specific impulse", the two parameters that must be maximized in order to get a respectable mass up to a high speed. So far, no one has ever done this: chemical propulsion gives good thrust, but a modest specific impulse (Saturn-Apollo V: heavy loads at low speeds), while, for example, an ion engine, provides a high-specific impulse but little thrust (small loads at high speed). With the energy efficiency of nuclear fission, we know that today we can obtain the winning combination of propelling large loads to high speeds.

The best nuclear way to boost a spacecraft to high speeds is by using the "Rubbia engine", powered with americium-242 or plutonium-239 as the fissile material. Both isotopes are "neutron hungry" and thus capable of absorbing the neutrons with which they are bombarded. The whole Am242/Pu239 process was studied in some detail by Carlo Rubbia and a band of enthusiastic amateurs from 1998 to 2002 at the Italian Space Agency, but it was later abandoned for no good reason, although we had come close to building a prototype, without having encountered any insurmountable difficulties. Certainly, though, we were still some way from building a real engine.

The actual engine, which would be capable of the right combination of thrust and specific impulse, is based on the heating of a gas (hydrogen) from the kinetic energy of fragments of nuclear fission. It seems difficult, but it is not. It begins with the fission of a heavy nucleus. When a nucleus of a heavy element (such as uranium-235 or plutonium isotopes, or americium and others) is bombarded with neutrons, a neutron can enter the nucleus and be absorbed. The energy deposited by the neutron into the nucleus "heats" it to the point that the core may rupture (hence the name "fission") into two or more fragments. Fission happens essentially because the nuclear forces that normally bind the atomic nucleus are no longer able to keep it together after it receives the energy input from the neutron. The fragments then fly off with a large amount of kinetic energy per unit mass, and this energy is usually deposited in the immediately surrounding material, i.e. in the same fissile mass, usually a bar of uranium, which consequently heats up.

Incidentally, nuclear fission was the physical principle that Fermi did not see, although it was he who had discovered the effect of bombarding nuclei with neutrons. It was demonstrated experimentally in 1938 in Germany and announced to the world, while Fermi, who had just received the Nobel Prize for his work on the neutron, was on the ship that brought him from Stockholm to New York and his new life in America. He recognized it immediately for what it was and blamed himself for what he had missed while at his laboratory in Via Panisperna in Rome.

Four years later, however, on December 2, 1942, Fermi had his revenge, with the discovery of the first controlled chain reaction, generated by the fission of uranium nuclei. It was the first atomic pile, in Chicago. The news of the success was communicated by Arthur Compton in Washington with a historic call made in improvised code: "The Italian navigator has landed us in the New World!" Perhaps like no other, Enrico Fermi was the explorer of new worlds, at least with his mind. Three years later, in 1945 in Los Alamos and at Alamogordo, Fermi was still the soul and brain behind the explosion of the first atomic bomb, which involves the instantaneous release of energy that occurs when you let the chain reaction run on unchecked.

In those final months of WWII, Von Braun, remember, was at that time still on the other side, in his missile centre in Peenemunde; being bombed, yes, but anyway a dangerous threat because V2s were still being launched one after the other. We should also keep in mind the following fact, especially after the famous letter of Einstein to President Roosevelt in 1939: that the Americans were not at all sure that von Braun was not preparing a V3, able to deliver an atomic bomb to London; and in Nazi Germany, there were some very good physicists who, after having discovered it, were still working on nuclear fission.

Let us return to our fragments of the heavy nuclei, broken up by fission. As we said, the fragments have a large kinetic energy per unit mass, which is precisely what makes them highly efficient energy packages. If fission takes place within a block of uranium, which is very dense, the fragments will travel for only a tiny fraction of a millimetre and will deposit their energy while colliding with the nuclei in their immediate surroundings. The result is that the material, the bar of uranium for example, heats up and this heat could then be transmitted to a fluid, for example water, as is done in a nuclear power plant.

In the Rubbia engine, the process differs somewhat. The fissile material is coated as a thin layer onto the inner walls of a hollow cylinder, a type of tube with a diameter of, for example, one metre and of 10 mt. in length. The tube is filled with gas: normally hydrogen at a pressure of one atmosphere. Thanks to this particular geometry, when fission occurs, there is a great probability that at least one of the (about two) emitted fragments enters the gas and deposits there its kinetic energy. Indeed, the thickness of the layer of fissile material, the tube dimensions and the pressure of the gas are all calculated precisely in order to optimize the best absorption of the kinetic energy of the fragments by the gas.

Evidently, at this point, the gas is heated, and all the more, the more efficient the fission process is. By carefully choosing the fissile material, its configuration, and how to bombard it with neutrons, etc., as well the overall geometry, it turns out that the central volume of the gas can be heated to the respectable temperature of 10,000 °C.

The tube in which the gas is brought to such a high temperature is in practice the central part of a rocket engine. All that is missing is a "nozzle", that is a steerable hole from which to let out the hot gas. In the nozzle, a small miracle of physics occurs: while the hot gas is inside the tube, the gas particles move in all directions, with a thermal motion which could not be more chaotic; but, as they are fed into the nozzle and exit from its open end, all particles by definition are forced to move in the same direction and so provide the needed thrust in the opposite direction. And *voilà*, it is done.

By dimensioning things carefully, this method of directly heating the gas can result in enormous savings on the amount of fissile material required. To get to Mars and back, just a few kilos of americium-242 or plutonium-239 is all that is required, which is equivalent to the volume contained by a couple of large glasses, since one litre of plutonium weighs about 20 kg. Other nuclear propulsion systems also exploit the heat generated by fission, which always means, ultimately, the kinetic energy of the fragments of nuclei that have undergone fission; but these other methods heat the gas to provide thrust in indirect ways, which are much less efficient and so require much more fissile material.

Of course, significant technical and structural challenges remain. For example, we do not yet know how to build a tube, much less a nozzle, capable of withstanding temperatures of many thousands of degrees. Moreover, because of the symmetry of the fission physics, the process described delivers more or less the same amount of energy to the walls of the cylinder as it delivers to the gas. There ensues, therefore, the tremendous problem of cooling the cylinder walls, which may be insoluble on Earth, but would certainly be more manageable in space. This means, for example, that the spacecraft could be fitted with large heat exchanging surfaces in contact with the "open sky", that is, with the rest of the universe, which is very cold at 2.7 degrees above absolute zero and also has an infinite capacity to absorb heat. Indeed, cooling the nuclear engine of a spaceship in orbit is much easier than cooling a nuclear reactor on the ground.

At this point, the engine is ready to be assembled and mounted on the high-speed spacecraft, the clipper that will carry the crew. For technical simplicity, and to avoid "political" problems, it should be assembled in orbit, so that the engine is never activated on the surface of the Earth, thus eliminating any risk of accident, nuclear or otherwise.

The various parts, each completely inert, would be transported from the Earth on board a chemically propelled type of Shuttle. Unfortunately, the place of assembly cannot be the ISS, for the reasons already mentioned: it was not designed for this and, above all, it is in a completely wrong orbit, too far to the north to get to Mars (it is a pity that the Soviet Union, at the time, did not extend further south: had Kazakhstan and Baikonur been on the equator, today's and tomorrow's space travel would be quite different).

While developing the engine on the ground, the spaceport/shipyard would be built in orbit. It would be best to put it in semi-deep space, say at an internal libration point between the Earth and the Moon. This is the legendary *"point neutre"* of Jules Verne, where the gravity of the orbiting Earth and Moon meet and "cancel" each other out. It is the right place to do build a base, both from the point of view of fuel consumption to get there, and because of ease of telecommunications. The structure that would be necessary to prepare for our spaceport should be very light, yet provide reasonably comfortable housing for the workers and engineers of the shipyard and one or more docks, i.e. spaces where the parts of spacecraft arriving from Earth would be assembled. How many shuttle flights to transport materials and how long it would take would depend only on the pace of the project. According to our preliminary calculations, the clipper may have a tonnage of around 150 tons and be the size of a modern passenger aircraft. It would be designed and built to be as light as possible, as fast as possible, and expose the crew to a minimum of risk, while providing good protection, both passive and active, to cosmic radiation. It would take about a dozen flights of a carrier of the Shuttle class (which had a bay load of 18 tons) to bring everything needed on site, and then, it might realistically take perhaps a couple of years to assemble the nuclear spaceship.

But for an expedition to make sense, there must be available in orbit around Mars, or perhaps already on the surface of Mars, much more in terms of stores and equipment than can be brought in the clipper. So, long before the manned ship at least two unmanned cargo vessels would need to be built, each having a goods capacity of 250–300 tons. These could be propelled chemically, so that they would be much slower than the clipper, not a problem if they started in good time.

A great deal of material would need to be shipped to Mars as cargo. For a start, the hydrogen fuel for the return trip of the clipper, which needs therefore to leave Earth with just enough fuel to reach Mars and then to brake for insertion into a Martian orbit, where the cargo vessel would be patiently waiting for it. The cargo vessel should also serve another shuttle-like vehicle (still to be invented, but chemically propelled) that would descend to the Martian surface and ascend back into orbit. The shuttle would need to be equipped with appropriate (and reliable) engines, fuel, and, above all, a landing system yet to be devised. I do not think von Braun's ski system will come back in fashion; instead, something better than retro-rocket braking, parachutes and airbags have to yet to be invented, possibly a gigantic sky-crane.

After choosing the landing site, much can be deposited on the Martian surface robotically, for example, one or more rover vehicles, with fuel and spare parts; and all the equipment for exploration in situ, from picks and shovels to geo-biochemical laboratories, with their building structures and furniture; and all the sealed bags and boxes in which to store samples to be returned to the Earth. All the necessary prefabricated building materials would be needed in order to prepare a shielded and may be pressurized shelter/laboratory. More supplies of air, food, water and something to read: for example, the book by von Braun: *Das Marsprojekt*. All these activities seem to be within the capabilities of the robotics that we already have today, let alone what we might have in the near future. In parallel, we will have to think about communicating with the Earth.

Today, and even more so in the future, the Pigafetta of the expedition would not be able wait until returning to Earth to give an account of his adventures, which in his case lasted more than twice as long as our planned adventure to Mars. Everyone will want to know everything right away, "live on live" as it were, even if Mars is very, very distant. But this could be done from the surface of Mars with a repeater, again sent by cargo and mounted near the shelter, to transmit the "diary" to one or more satellites in orbit and then to the cargo vessels and to Earth.

Telecommunication satellites, which by then will already be orbiting in sufficient numbers to ensure continuous coverage, and the cargo vessels, would each be equipped with adequate antennas and transmitters. In addition to the transmission to Earth, we must also ensure that dialogue between individuals, vehicles, etc., will possible on the surface of the red planet. The scouts will certainly have much to say and will do it better than the commanders of the ships of Columbus or Magellan or Drake who could only wave flags when they were within sight of each other. Above all, there would need to be an emergency channel in the event, for example, of a Martian sandstorm arriving or, worse, of a solar particle storm. It will be possible for the alerts to such dangerous events to be sent in time, since they will have been seen by observers on board the ships remaining in orbit in contact with Earth observatories.

Upon receipt of the confirmation of the safe arrival of all the necessary material in Martian orbit and on the surface, the clipper, with its crew of 4–5 people, would leave Earth, choosing the time of year when the relative positions of the two planets was appropriate. The crew would travel to the spaceport at the libration point by a chemically powered shuttle, arriving "in shirtsleeves" as relaxed passengers of a transport driven by others. After the last checks, handshakes and hugs, the voyage to Mars begins, following a segment of an ellipse calculated to reach the orbit of Mars in the exact spot where the planet will be upon arrival. In short, you have to arrive in the right place at the right time, because Mars travels in its orbit and does not stop nor turn back for late arrivals.

It will take exactly 150 days to get to Mars. Once there, the manned spacecraft would brake for insertion into Martian orbit and then dock with the cargo vessel, which would have been spotted (with some relief) just beforehand. It is one thing to hear a radio signal or see a blip on a radar screen; it is another to see with one's own eyes the ship that contains, among other things, the fuel to get home. At this point, while one or two crewmembers remain in orbit to control the supplies, maintain the clipper for the return journey, peer down to the surface of the planet from above and act as a central communication point with the Earth, the other lucky crewmembers enter the landing vehicle and land on Mars.

The descent is a very critical event: on Mars, as we have said, tremendous and unpredictable sandstorms can occur. Despite the thin atmospheric layer and its low density, these storms are weather events that can completely obscure the planet and can be lethal to a spaceship that has to land in exactly the right place, where robots have left a guiding beacon and survival equipment. Fortunately, the descent should not take long, maybe an hour, and one can easily check from orbit if there is a storm coming. The risk seems acceptable. After landing successfully, it is critical to check that the spacecraft's descent manoeuvre has left it parked in a comfortable niche, with no dangerous inclinations or any unstable rocks under its feet (remember the Apollo LEM…). It is important to consider checking out at once the vertical launch system towards the orbiting clipper, waiting to take us back home in exactly 41 days. Only after being sure that the descent vehicle is parked well and ready to go back up can we relax and look around.

Incidentally, an interesting place to use as a landing runway (and maybe even as an ice skating rink) could be the crater created by the impact of a comet with the Martian soil. The body of the comet made of ice, melted on impact and then solidified into a beautiful ice lake inside the crater. The size of these cometary lakes can be several hundred metres in diameter, and many tens of metres thick. If it were possible to measure their strength, i.e., their ability to withstand a landing, they could really be an ideal landing place since they are flat, with good visibility, and with the edge of the crater as shelter from the furious winds. They would also allow for a good anchorage, as they are formed of a solid material whose properties we know well. They would be especially useful if a first chemical analysis would show that the ice could be turned into drinkable water. If not drinkable, it would still have many uses, not least the production of oxygen, given sufficient energy.

More generally, the discovery of solid water on the surface, or even liquid water at accessible depths, could be a determining factor in deciding where to send our Mars Lander. For this reason, future robotic missions to Mars, after Curiosity, will increasingly be dedicated to drilling, for example, with core drills to search for any available water. We have to hope to be lucky, or to be able to send a robotic version of a dowser's wand to Mars.

Survival on Mars, especially the first time, will not be like a holiday in the Maldives. Apart from the lack of any ocean, the thin and rarefied atmosphere (equivalent to being at twice the height of Everest), as its composition of 95 % carbon dioxide, and the very low temperatures that undergo extreme changes, will force us to forget swimming costumes and sun lotions, and to don instead an uncomfortable spacesuit. It will be a modern version of those Apollo spacesuits, of course, but the physics of the difference in pressure and temperature still necessitates something very strong and stiff, which is particularly painful on the hands. Today, the hands are always the weak point for astronauts doing spacewalks in their spacesuits, because they need to be able to move continuously against the internal pressure, but they have relatively weak muscles. On the journey from Earth to Mars, our heroes will have to work at a special form of gymnastics to strengthen their fingers in order to be able to bend the stiff gloves of their spacesuits without having to stop to rest every few hours.

Hopefully, in the meantime, we will have managed to produce a kind of pressurized and servo-assisted exoskeleton, that is, a spacesuit which is tight and well insulated, but also lightweight and flexible enough to allow the astronaut free movement of all the joints of the human body, including the hands. The explorers of Mars will have been able to abandon the "Michelin Man" model, which has remained essentially unchanged from the Apollo missions to the extravehicular activity of the space station of today.

An important problem will be a source of energy for the exoskeleton. We could cover the helmet and suit with flexible solar panels, but we would not go very far on Mars, where the sun is quite pale. No, for explorers on Mars we need to develop new forms of electric accumulators, perhaps derived from the recent successes of organometallic chemistry, for example, those based on clusters of platinum atoms surrounded by clouds of carbon atoms. This really is a nanotechnology currently being studied in Italy at the University of Milan, which has a configuration that seems to be very promising for the future extremely high-density accumulation of electric charge.

Another detail that needs to be considered very carefully, and of which little is said, is the problem of contamination with Martian dust, which is present everywhere and is constantly being blown about by the wind (even without reaching the levels of real storms, which can anyway last several days or even months). Not only could the Martian dust be abrasive to the exoskeleton, but it may also be irritating or toxic to the human body because of its high content of oxides, and possibly toxic, due to the presence of undoubtedly carcinogenic hexavalent chromium ions. In short, we should somehow remove and leave the dusty exoskeleton "out of doors", which will not be easy, or leave it in a decontamination chamber, which is yet one more complication to think about. The Apollo astronauts had not cared a lot about safety, when getting back into the LEM wearing their spacesuits dirty with lunar dust, electrostatically stuck to them and which undoubtedly spread around the environment in which they had to stay for a while longer—but no one seems to have suffered. At the beginning, our explorers should ensure, just as Marco Polo did in the desert, or Amundsen at the pole, or Hillary on Everest, that their shelter works well. It will be more than just a tent: it must be a solid, windproof structure, giving protection from radiation and meteors, and be pressurized and heated. It is not a trivial problem. So far we only have abstract ideas on how clever robots could overcome these problems.

Our human explorers, however, will spend most of their time exposed to outdoor hazards, taking risks in the open where there is a whole new world to discover. They will naturally have a detailed programme, which would have been the subject of endless discussions between planetary geologists, chemists, biologists, etc. But, as always, in the field detailed plans can only be approximated, since the conditions on the ground will certainly differ from expectations. Which is precisely why it is necessary to send humans to explore: their intuition, flexibility, ability to improvise and their courage are all qualities that are more human than robotic. One certainly needs humans (both men and women), who will have as solid and as broad backgrounds as possible in terms of science, culture and technical skills: they must be able to recognize materials and conditions, but equally well they must not get into a panic if, for example, the rover develops a fault, or if communications are suddenly interrupted.

Let us use our imagination as to what might happen on the surface of Mars. For example, what if our heroes, wandering among the reddish, rocky escarpments suddenly see in the distance a figure in a human form. On approaching it, they realize that it is a very well executed statue of a real human being. It is of a gentleman with moustache and goatee, wearing rather old-fashioned clothes, with a telescope in one hand and a piece of paper in the other. At the base of the statue, in Latin characters still clearly legible, there are three words: "GIOVANNI VIRGINIO SCHIAPARELLI".[1]

1 Giovanni Virginio Schiaparelli (1835–1910) was a great astronomer and a prolific writer, who was the director of the Observatory of Brera in Milan for 38 years, from 1862 to 1900. In 1877, he began a long series of observations of Mars, and it was he who suggested the presence of the famous "canals" on the Martian surface. Through them, following the Martian seasons, he imagined, water flowing from the seas and oceans of the planet with the melting of the polar ice caps. As well as being a distinguished scientist, Schiaparelli explored with the mind, sometimes more than with the telescope, and was also an excellent dreamer. The passage of his, reported below, was published in 1895, in the journal "Nature and Art", with the annotation in pen *semel in anno licet insanire*: once a year you can be insane…

Thanks to the techniques used by the unknown Martian sculptor to finely incise the etched inscription, as our scouts in their exoskeletons approach the statue with caution, they are able to read the passage below, which is clearly written on the paper that Schiaparelli holds in his hand (the scouts are educated, so maybe they can read Italian; otherwise, they could have it instantly translated by the facilities offered by the solar-system-wide web that they have in their helmets). Here is the text, reported by the unknown Martian artist:

> It will not be difficult to imagine designing the works needed to regulate the distribution of water according to where and when it might be needed, throughout all the irrigable territory. Mars must certainly be a paradise for plumbers! And regarding social order, Mars could also become a heaven for socialists – a place where international disagreements and wars are certainly unknown, and where the enormous amount of study and work and effort, which the insane inhabitants of another nearby globe consume in harming each other, is instead all directed to fight the common enemy, namely the difficulties that Nature presents to oppose our each and every step.

Somewhat confused and very excited by these unexpected words, our scouts take a good look around, searching for plumbers (socialist or not), but at the moment they do not seem to be around on the dry Martian surface. What a pity; because there is that dripping tap on the spaceship that needed fixing… In their helmets, they listen to the harsh and futile discussions that are taking place on Earth on the gnostic–philosophical–theological consequences of their discovery, which are, of course, being broadcast live but that, on the red planet, really seem a long, long, way away.

After a little more than a month of experiencing emotions like this, and other adventures even more difficult to recount, it is time to prepare to leave the planet and begin the long journey home. As for the samples to be taken back to Earth, there will be space for more or less one ton of various rocks, which are now carefully sealed and protected. Everything that does not serve a use, including the descent engine, is abandoned and left on the face of the virgin Martian soil as pollution. Too bad. The shuttle will leave at the right time for a low orbit (say 200–300 km) rendezvous with the clipper, which would still be attached to the cargo supply vessel. After being fully refuelled and provisioned with air, etc., and loaded with the precious Martian samples, the clipper is undocked, released from all that is destined to remain in Martian orbit (and which may be useful to another expedition) and proceeds to reignite (fingers crossed) the main nuclear engine.

The departure from Mars after 191 days of the mission is also designed to allow insertion into an Earth re-entry orbit as soon as possible. Indeed, this time the trajectory involves a shortcut in the Solar System, exploiting the relative motions of the Earth, Mars and Venus. The shortcut goes inside the orbit of the Earth, with a close fly-by of Venus to get a gravitational slingshot effect. All this should be done with caution, however: if the Sun seems pale from Mars, near Venus solar radiation is about double that on Earth. The joke of the astronauts is all too easy: more than a "Venus fly-by", it could become a "Venus fry-by". After the 300th day of the mission, the worst is over. Inside the clipper, everything begins to look too cramped: the crew think about a dip in a pool of warm, liquid water, or a terrestrial toilet (two situations where gravity plays an important role), or lying down to sleep in a bed, maybe with a partner (where gravity is also helpful…). Meanwhile, the Earth is seen to be coming ever closer, until at last the small, grey Moon appears next to it, and then the bright spot close by, which is our goal.

Exactly one year after starting out, with the Earth-Moon system in the same position relative to the Sun as when we left, the deceleration begins in preparation for docking with the spaceport between Earth and the Moon. If everything has gone as it should, and if we had correctly calculated the right orbits and fuel consumption, etc., and accounted well for every little supply that we held, we should arrive and berth on the 369th day since departure. There is now just time to give those at the spaceport a hug: after all it is them who built for us the clipper, which was our home for almost a year. And they built it very well, it seems. After a clean-up, and the inevitable change of some parts, such as the nuclear engine, the clipper will be ready for another trip.

The chemical shuttle is ready to take us, with our load of rocks and memories from another world, back down to the Earth. A few more hours and then, finally, there is gravity, the blue sky, a cold beer, a beloved face to look upon with tears in one's eyes, although seen from behind the glass walls of the quarantine quarters in which we will be locked up. Our muscles, bones and head need a lot of recuperation after a year in zero or very low gravity and of always being confined in restricted spaces, outside of which was an environment that would not forgive the slightest mistake. We are tired, but it was worth it.

The cost of this first visit to Mars, billed to all mankind, is a trillion dollars, or at most two, including research and development, construction, testing and the return mission. All was done over a period of ten to twenty years. Of course, money could have been saved by avoiding the return journey. This may sound like a joke, but it is not. An eccentric Dutch entrepreneur has already collected thousands of applications from people willing to volunteer going to Mars on a one-way ticket: one goes, colonizes and remains there, the others arriving in successive waves. We make no comment, preferring instead to present a more realistic case, with the cost of mission standing at a mere trillion dollars, including the return. By way of comparison, note that according to the 2011 report by SIPRI (Stockholm International Peace Research Institute), the total military spending worldwide is $1.74 trillion a year. In 2011 alone, The USA spent $711 billion (about 41 % of the total), three to four times the cost of the entire International Space Station, with its shuttle, a project that lasted for thirty years. Second in the league for military spending is China, with $143 billion (soaring); Russia is third with $72 billion (also increasing sharply), then the UK and France on a par at $62.5 billion, and a few others do the lion's share of the rest: Japan, India, Saudi Arabia, etc. Italy has officially spent $34.4 billion, but according to SIPRI that number is "less than transparent". And it is fortunate that by now it is more than twenty years since the Cold War between the superpowers… I wonder what the spending would be now if that were not the case. Von Braun was definitely right: a project lasting at least ten years for a mission to Mars would cost about the same as one year of spending on the military. But, after the first visit, we will definitely return, using everything we would have learned and taking advantage of all the materials that would have been left on the surface of Mars and in orbit. However, on that next time, we would want to make ourselves a semi-permanent base on Mars, not just an emergency shelter. The experience gained on the Moon will be very helpful in finding the right underground place in which to build a spacious, solid, secure and pressurized base.

Perhaps on Mars too there are lava tunnels, like those we went looking for on the Moon; or at least volcanic cavities, which may have a little residual heat and perhaps underground sources of liquid water. Unfortunately, in the absence so far of knowing anything for certain, we are bordering on the realm of dreams. Maybe it will take one (or more) exploratory missions to find a site suitable to serve as our second underground home on Mars. In his fantastic book of technical science fiction, von Braun had already thought of an underground home, and more, on Mars. His explorers land on skis and then move about in tanks (tradition counts) on the surface, where they come across strange but clearly artificial structures that protrude from the ground. Approaching with caution, they discover that there are Martians living inside them. They are more or less similar to us, but smaller and darker; in short, they are a slightly inferior race. However, our tall and white explorer quickly makes contact with the natives and almost immediately understands their language, although it is not clear how this is done. The visitors from Earth, or at least the head of the expedition and his senior officers, are immediately given a tour of the subterranean civilization and shown the Martian, cities linked by an excellent, pressurized subway system. We would have a lot to learn from von Braun's Martians.

But reality is often stranger than fiction: there are those, such as NASA, who are designing even longer term, and much more grandiose plans. Some people even think of radically transforming Mars, making it turn back in time to a period when liquid water flowed on the surface and there was a reasonable atmosphere, which is necessary just to keep it liquid. This means trying to make Mars similar to the Earth, to "terraform" the planet. Given an atmosphere and liquid water, it could become a feasible undertaking, if done on a geological timescale, of course. It would still be a fascinating project and something that would require vision and optimism.

We have no idea how to transform the atmosphere of Mars, nor, admittedly, do we know how to find sufficient underground water or the almost endless energy reserves that would be needed. A large proportion of the carbon dioxide and water vapour that once formed the atmosphere of Mars is now trapped in the planet's surface, especially in the vicinity of the polar caps. If it were possible to warm up areas of the planet covered with permafrost, perhaps by concentrating sunlight, it might be possible to release CO_2 and water as vapour to form a slightly thicker atmosphere than the present one. This would be an ideal starting point to trigger a greenhouse effect, which would allow the warming process to become self-sustaining. Remember that on planet Earth, it is exactly the greenhouse effect in our atmosphere (which is over a hundred times thicker than that of Mars) that allows the global average temperature to be around 14 °C, and thus allows water to remain liquid and, ultimately, allows life to exist on Earth. The greenhouse effect is, in short, something which you should not play about with: if it is too exaggerated, as on Venus, it prevents life because it is too hot; and if it is too feeble, as on Mars, life may either never begin or it may be extinguished.

Reforming the atmosphere that Mars has gradually lost, over one or two billions of years, with the creation of the "correct" level of greenhouse effect would allow the surface temperature to return to acceptable levels (for humans), and water to the liquid state. Of course, at some point it might be considered that free oxygen should be released into the Martian atmosphere, since this is what we breathe on Earth. But, just as occurs on Earth, oxygen must continually be created or re-created, because it combines rapidly with many elements of the Earth's surface. A start could perhaps be made by bringing some robust forms of terrestrial bacteria to Mars, such as those that live quite happily in the icy deserts of Antarctica, and see if they might take root, maybe in some protected, irrigable area. Then, gradually, more and more abundant plant life might evolve. After all, it is thanks to the first forms of life that free oxygen entered the Earth's atmosphere; and it is thanks to the plants and forests that today we are able to breathe, at least until we destroy them all, and with them our oxygen.

It is hard to say how long it would take to restore the atmosphere of Mars, or to make it similar to that of Earth. Perhaps only at that time would we be able to explore it more calmly. It would probably take many thousands of years, if not millions. But then we already know that any atmosphere will not last forever, because the planet has already lost one. However, it could perhaps last for a billion years. Who knows, maybe there is an indirect form of exploration yet to be invented.

Transforming Mars using terrestrial bacteria, in the absence of any appropriate Martian species, as well as being in itself an operation so complex and expensive as to be at the limits of science fiction, it also seems even more questionable in other ways. The export of micro-organisms from the Earth to Mars would be a colonization worthy of the worst violence of that dark age of colonialism that occurred between 1500 and 1900, with the cancellation of one planet's individuality in favour of another's. Frankly, I would find it unacceptable.

Already, however, NASA has given us some certainty concerning Mars: the chemical composition of Martian soil is particularly suitable for the indoor cultivation of asparagus. Yes, asparagus: it thrives in sandy soils, rich in iron and other minerals, and it will require only a little water and fertilizer to transform underground tunnels into places for the cultivation of those species of asparagus, and perhaps mushrooms too, that grow well in the dark. They would need to be consumed locally, though: the transport costs would be too much for even the finest restaurants on Earth… but who knows, a plate of Martian vegetables could still find its wealthy terrestrial admirers.

Chapter 8
Sphere +3 Outer (or the Solar System)

Homo sapiens and his Machines in the (Rest of the) Solar System

Breaking through Sphere +2 and going beyond the Moon sending probes to the planets, starting with Mars, is something that not even Jules Verne imagined or wrote about. Instead, the proof that reality always beats imagination was already contained in the programmes and capabilities of the very first generation of space explorers, from Korolyov with his Soviet school, to that of von Braun and Carl Sagan at NASA.

After reaching the Moon with his unmanned Luna probes, Korolyov realized the possibility that his rockets could go much further. Immediately aiming for Mars, he came up against a number of real problems of planetary navigation; real since this was much more than the simple case of going from the Earth to the Moon. It required not only adequate power for launching, but much more sophisticated instrumentation than was available in the USSR in the late 1950s. From 1960 to 1962, the Soviets launched at least five probes to Mars, all of which were lost.

Meanwhile, von Braun and his team were not quite sitting on their hands: the Jupiter rocket was abandoned in favour of the new and more powerful Atlas–Agena, which was used, starting in 1962, to launch a programme of planetary exploration with the Mariner probes. It is important, historically and politically, to note that the Mariner programme went on for more than 10 years, until 1974, in parallel with the Apollo programme, which, absorbing as it did the bulk of the NASA budget, grew dramatically during the 1960s, when the Vietnam war cost was also ramping up.

The first Mariner was destroyed shortly after take-off; the second passed close enough to Venus to be considered a success; the third was also lost. It was Mariner 4 in 1965 that rewarded the tenacity and boldness of NASA. Launched with the Model D Atlas–Agena, then the most powerful rocket in use because the famous Saturn V was still under construction, Mariner 4 passed within a few thousand kilometres of the red planet and took 22 pictures of its surface, successfully sending the images back to Earth.

These 22 photographs taken by Mariner 4 marked another milestone in the long history of exploration by *Homo sapiens*. It is difficult to make comparisons with other moments of exploration. The first pictures of Earth from space? The first

© Springer International Publishing Switzerland 2015
G.F. Bignami, *The Mystery of the Seven Spheres*,
DOI 10.1007/978-3-319-17004-6_8

photograph of the South Pole? The recounts of Pigafetta? "Il Milione" of Marco Polo? It is futile to look for comparisons. But what these photographs of Mars actually did was simply to close the door forever on Schiaparelli and Lowell's dreams of canals and an inhabited world. We discovered that Mars is another planet, with no buildings and no princesses of Edgar Rice Burroughs, the American writer who in 1912, two years before inventing Tarzan and Jane, had inaugurated a series of adventures on Mars. The reality of exploration once again exceeded imagination (sorry, Jules; and you too, Wernher). It is likely that Korolyov, who died in January 1966, never saw those pictures; pictures he himself would have desperately liked to have taken with probes of his own: maybe it is better that way.

The same year, however, scientists from the Korolyov School attained another beautiful record: on February 3, 1966, the 9th in the Luna series performed the first soft landing on the lunar surface and took some exceptionally sharp photographs, the first that were ever taken from the surface of a celestial body other than the Earth. The 1960s were exciting times for space exploration! Incredibly, but to the eternal credit of NASA, the Mariner series also continued in 1969, alongside the great Apollo project. Mariners 6 and 7 successfully passed close to Mars and sent back an even greater amount of beautiful photographs in early August, which went rather unnoticed arriving as they did just after the legendary Moon landing on July, 20th. After Mariners 4 and 5, a second-generation vehicle was used to launch Mariners 6 and 7: the new combined Atlas–Centaur rocket. This was almost as powerful as the Saturn V, which was then entirely reserved for the Apollo project.

On the strength of these further photographs from Mariner, August, 4 1969 found Wernher von Braun, the hero of the moment, appearing before the Commission of the US Space Agency. His speech is still on record, of course, and in it he more or less said:

> By reaching the Moon we have demonstrated that we know how to do it and that we have kept the promise made to our assassinated President; I now propose a manned expedition to Mars.

"Dear Wernher, I still mention that presentation of yours … do you remember?" I say to that distinguished gentleman, who has suddenly appeared in front of me at the table as I write. He has materialized on his own initiative, rather intrusively, and without even my needing to skim a book! He is very interested to know how space exploration is going on now, and how it could possibly manage without him.

Born in 1912 (like my father), he is now an overcentenarian, but continues to have an erect posture, a slight military bearing and a heavy German accent. He has not lost his authoritative tone of command and is a little arrogant.

"*Sicher*, of course I remember it, *zum Teufel*, damn it, we lost Mars for very few votes …" This dialogue, I say to myself, is better than an impossible interview. I have here von Braun in person and I can ask him whatever I want.

In 1976, two identical probes, the Viking missions, descended to the red planet's surface, in search for life. The distance that separates the sphere of the "Heaven of Earth" (Sphere +1) from that of the Moon (Sphere +2) increases by a factor of about a thousand when we move from the Moon to Mars. There is a harmony, or at least a periodic symmetry, in the scales of space and time among these spheres of exploration.

Today, we have almost lost count of how many missions planet Earth has sent to Mars. Perhaps forty, including those sent by the Russians, Americans, Europeans and the Japanese. About half have failed, with a failure rate that, surprisingly, does not decrease much with time. An indication, perhaps, that ever more ambitious missions naturally become increasingly complex and thus more prone to failure. Less than half of the successful missions to Mars also had more or less good fortune with their surface landing. The first, again, were the Russians, in 1971.

"*Genau*, to be exact, the Russians were the first to land softly on Mars, even if their instrument, Mars 3, worked for only a few seconds" notes Wernher with just a hint of bitterness.

On the Martian surface, today almost ten tons of scrap metal from the Earth have been left behind. We hope that it was all really sterile junk, but unfortunately, we do not believe that to be the case. There exists, and there above all has existed since the first robotic exploration of Mars and the planets by *Homo sapiens*, the serious problem of contamination. It is perfectly possible that we will eventually discover life on Mars, only to realize that it is bacteria of Earthly origin left on a probe from a technician who sneezed while assembling it. This may sound like a joke, but it has already happened in the case of the Moon, as von Braun knows very well (he chuckles). We have not yet found life on Mars, although if we were to really begin looking for it, who knows...

Our reward for having placed so many probes in orbit around the red planet is that we may now know the surface of Mars almost better than that of the Earth. In the sense that we have terabytes and terabytes of data that show every detail in images of Mars, square metre after square metre, aided by the excellent transparency (in the absence of sandstorms) of the Martian atmosphere, which is less than one-hundredth as dense as ours. We see ice avalanches when they fall down the cliffs near the poles in the spring; we see dunes that have moved after sandstorms; and we note annual differences in phenomena that are probably due to the flow of liquid water. We have seen the formation of new small impact craters, and we have already found, and photographed, meteorites fallen on the surface of Mars. The names, in large part, have already been given: Google Mars is a reality, after Google Earth.

And then, in the last fifteen years, NASA has been able to send to Mars three successive generations of mobile robots, each able to explore the surface in greater detail and able to move further and further afield. They are led, by observers/drivers on Earth, to places judged to be most interesting for a detailed micro-exploration. The latest robot explorer on Mars, Curiosity, has a name that says it all. It is already able to wander among stones and dry mud, in what looks like the bottom of a lake now dried up because any liquid water has all evaporated. A place chosen in the hope of turning a rock over and perhaps find underneath it a fossilized fish to photograph ... It is possible, but it has not happened yet. Who knows? Certainly, like the two Viking probes, Curiosity has the instrumentation necessary to find a signature of biological activity, although it would be of a form of life much more basic than a fish. As, indeed, was all elementary life on Earth for 3 billion years after it first appeared.

With Curiosity, we can now begin to understand thoroughly the composition of the soil on the Martian surface and the presence of organic molecules or biological compounds among the rocks. The hope is to find a good scent track and then follow it by "sniffing" (which Curiosity does particularly well with its vapour analysers) and so eventually come across its source. Overall, from 1971 to today, in less than half a century, *Homo sapiens planetarius* has not done a bad job in space exploration directly on the surface of another planet. Giovanni Schiaparelli and Percival Lowell, who made their fundamental contributions a little over a century ago, would ask us: what remains to be explored today on Mars? Well, the best is yet to come.

The interior of Mars, for example, is something about which we know very little at present. With seismographs on the surface, we could explore it "indirectly" as has been done to a great extent on Earth and to a certain degree on the Moon as a by-product of the Apollo programme. Who knows what the interior or Mars is really like! It probably does not have a significant liquid core, since it does not have a magnetic field on a large scale, i.e. aligned between the North and South Poles, as on Earth. If it had one once, then it must have slowly dissipated, probably as the small planet cooled. Understanding how this might have happened may be very important to our understanding of the future evolution of our own magnetic field on Earth. Although we know it to be a bit fickle, so far, over the last 4.5 billion years at least, it has fortunately not yet let us down.

On something like the existence of an ordered magnetic field on our planet, we should remember, the future of the human species really hangs in the balance. Without its magnetosphere, the Earth would be like Mars: there would no longer be a shield to ionizing radiation coming from space and there would follow an almost perfect sterilization of the planetary surface. That is why we expect some of the most interesting results concerning Mars, and the possibility of biological activity on it, not to come from Curiosity (though some surprises may still be possible), but from another robotic explorer, planned in the near future by ESA with a particularly significant Italian participation: ExoMars. Within a few years, it will be able to add a special dimension to the exploration of Mars by investigating its underground environment, or at least the Martian environment just below the surface. ExoMars will be able to drill into the Martian soil, from dry mud and sands and other stuff up to very hard basalt. It will do so through a special drill probe, equipped with a small imaging spectrometer, constructed in Italy. It will be capable of penetrating up to two metres below the surface, where, potentially, the conditions could have been, and maybe still are, much more favourable for the development or preservation of life. And we hope that the probe that will bring samples to the surface will have been properly sterilized.

And what about Venus and Mercury? We will mention them only briefly. They can be explored, and in part, they already have been explored and could still be, with dedicated robotic probes, or during a flyby of a manned mission to Mars. As is known, the environmental conditions of our two sister planets, that is those in orbit between us and the Sun, are absolutely prohibitive for manned exploration, because of their surface temperatures. In the case of Venus, moreover, the atmospheric pressure at the surface is comparable to that of a super-planet. It is almost one hundred kilograms per cm², which equivalent to being, on Earth, at a depth of thousand metres under the sea.

No, we have decided that our two inner planets are certainly interesting to explore in order to understand the origin of the solar system and the distribution of matter in the original proto-planetary disc, but they are not of burning interest for human exploration. Let us go back out beyond Mars, where we still have left to explore most of Sphere +3, the one that will take us right to the edge of our Solar System.

So far, few probes have been sent out beyond Mars. There have been enough to give us an idea of the immense possibilities offered by the exploration of the outer Solar System, that is, the immense space that extends out beyond the four rocky planets: Mercury, Venus, Earth and Mars. Here, I realize that I need to briefly acquaint my "friend" Wernher on what has happened in the last thirty years in this part of the heavens, since he has ceased to deal with it, carried away as he was by an incurable disease in 1977. And above all, I would like to hear his opinion concerning the future of this area of exploration. But first, here is my summary of all man-made objects that have been gone beyond Mars.

The two probes, *Pioneer 10* and *Pioneer 11*, launched by NASA in 1972 and 1973, respectively, to traverse the solar system.

Two *Voyager* spacecrafts launched by NASA in 1977 to traverse the Solar System two years after the two great unmanned Viking missions were sent to explore Mars.

The large NASA *Galileo* probe launched from the Shuttle in 1989 to explore the Jovian system.

The NASA-ESA *Ulysses* probe launched by the Shuttle in 1990 to explore interplanetary space from a point near Jupiter. It exited the plane of the ecliptic, the first and only man-made object to leave that plane, where we and all the planets exist.

The NASA probe *NEAR-Shoemaker* launched in 1996 to explore the asteroid belt beyond Mars, in particular the objects Mathilde and Eros, on which *NEAR-Shoemaker* still lies, after having taken close-up photographs.

The large NASA-ESA *Cassini-Huygens* probe launched in 1997 to explore the Saturn–Titan system, with the European probe landing on Titan, the most distant celestial body on which a human artefact has ever landed.

The NASA *Stardust* probe launched in 1999 to explore comets and other objects (touching, for example, the asteroid Anne Frank): it returned to Earth in 2006 with a handful of dust from the comet Wild 2 and other dust coming from outside the Solar System.

The Japanese probe *Hayabusa* launched in 2003 to explore the asteroid Itokawa that was able to land there and after an incredible adventure returned to Earth in 2010, bringing with it a pinch of dust from Itokawa.

The European probe *Rosetta* launched in 2004 by ESA to rendezvous with the comet Churyumov–Gerasimenko in the outer parts of the solar system, away from the influence of the Sun. This took place in 2014 and the probe's lander *Philae* successfully landed on the comet, but in a shaded hollow where it is currently in hibernation.

The probe *Deep Impact* launched by NASA in 2005 to study the composition of comet Temple 1 by launching a projectile to impact on the surface, after which it went on to make observations of exoplanets and make a subsequent photographic visit to another comet, Hartley 2, an encounter which occurred in 2010.

The NASA *New Horizons* probe launched by NASA in 2006, which passed Jupiter in 2007 with the ambitious goal of reaching Pluto, and its satellite Charon, in 2015. A particularly touching or macabre aspect is that it carries on board a small amount of the cremated remains of Clyde Tombaugh, the American astronomer who discovered Pluto in 1930.

The NASA *Dawn probe* launched by NASA in 2007, which visited the asteroid Vesta in 2011–2012, and is now flying to Ceres, the first asteroid to be discovered by Cesare Piazzi, director of the Observatory of Palermo, in 1801. It arrives at Ceres in 2015.

This makes about 14 man-made objects that were designed to go beyond the orbit of Mars, to explore what, for us, is the outer part of the Solar System. Technically, to this list could be added objects that have gone beyond the orbit of Mars "involuntarily", such as members of the Mariner series, which were designed to photograph Mars in passing and then disappear into deep space. But these do not count as real explorers of Sphere +3 beyond Mars, as they are only objects that ended up in a ballistic flight after finishing the task for which they had been designed.

"*Wunderbar*, thank you for this nice little summary." says my friend Wernher, happily. "I see that since I got out of the way, NASA has given itself much to do ... but why have no humans yet gone beyond the Moon?"

"Yes, I must admit, we have not yet succeeded there ... To go beyond the Moon, to Mars, or even further, to the outer parts of the Solar System, we have not had the courage to make serious plans ..."

Clearly annoyed, von Braun takes himself to an imaginary blackboard and begins to write with his finger. The substance of his speech is clear. He had always known that to go beyond the Moon, for example just to Mars, it would require energy from nuclear fission; a form of energy he knew and understood well, since they spoke about it in Germany during the Second World War.

But now Wernher knows that going beyond Mars would take a quantum leap in the amount of energy used. He therefore proposes to use instead a method based on nuclear fusion for spacecraft propulsion. This is a process where, per unit mass of fuel (i.e. a proton, for instance), the energy released is much greater (by a factor of five to ten) than what is released during fission. A spacecraft powered by nuclear fusion could be a huge step forward for the human exploration of Sphere +3, with visits beyond Mars, to the asteroid belt or by intermediate steps (and landings?) to the large rocky satellites of the gas giants Jupiter and Saturn. Only that, unlike a fission-powered spaceship, we do not yet know how to build one powered by fusion.

We know very well, first from astrophysics and then from theoretical principles and practical engineering work done both in military and in civil fields, the basic physics of fusion. Simplifying the story a lot, let us say that it involves bringing into close proximity two protons, for example, or two nuclei of the hydrogen atom. Because both carry a positive charge, it is very difficult to bring them close together. But if one succeeds, another force suddenly comes into play—the strong nuclear force—that acts like a special glue to unite them for life. If deuterium or tritium is used, i.e. hydrogen isotopes in which the proton is tied to one or two neutrons, fusion can result in the formation of a new element, helium, just as it happens in the stars. And, as in the stars, physics says that for each fusion of two elementary particles, a large amount of energy is liberated, greater than what is released in nuclear fission, the physical process in which, instead, the nuclei are split apart. In short, the explosive release of energy is potentially a bomb ... and in fact, as soon as we understood the process and learned the technology, the first thing for which *Homo sapiens* used the physics of fusion was a bomb, the so-called hydrogen bomb, or H-bomb, so powerful that it was, and so far is, the most powerful and destructive weapon ever designed and built by man.

To build a bomb is relatively easy. The conditions in which fusion can be initiated are obtained by the explosion of a classical fission bomb, based on uranium or plutonium such as those used on Hiroshima and Nagasaki. The explosion of a fission bomb briefly generates the extreme conditions of temperature and pressure necessary and which last long enough to allow the fusion of hydrogen nuclei. The difficult part is positioning the hydrogen nuclei with the necessary accuracy. When, triggered by the first bomb, the right conditions are met, and the second, much more powerful process occurs, which is the (uncontrolled) fusion of nuclei.

What we would like to do with the fusion process, but are still unable to do, is the equivalent to what Fermi's atomic pile does for the fission process. That is, for a slow and controllable reaction to take place, i.e. what Fermi managed to do with his brilliant idea to moderate the neutron flux responsible for initiating the chain reaction. We do not know, in the case of fusion, how to get the reaction to take place in a slow and controlled manner, but at the same time to be sufficiently sustained to give a continuous net flow of energy. We have been working for decades on the problem; perhaps it will take more decades until we succeed. Of course, when we will successfully master fusion in the laboratory, a fusion engine for spaceships will be, at least conceptually, within our reach, just as fission-powered spacecraft is within our reach now.

The way to achieve controlled fusion, whether on the ground or in space, must involve keeping the fuel (hydrogen or its isotopes, such as tritium and deuterium, or helium) within a magnetic container, where one can exploit the fact that the proton is charged. Once contained, the temperature of the trapped particles must be increased. And therein lies the difficulty—how to build a "magnetic bottle" to contain the gas, which at these temperatures is in fact called a "plasma", heat it up to ignition temperature and then exploit its release of energy. Of course, building a "magnetic bottle" capable of triggering and sustaining fusion and capable of being installed on board a spaceship is as impressive a technological and scientific challenge as they come. Yet, there are those who will try, even if it seems impossible.

"And fortunately there are people who try. One of my favourite sayings was: I have learned to use the word impossible with the greatest caution." says Wernher, gravely. He adds: "Only by a miracle could the scientists of the eighteenth century have foreseen the birth of electrical engineering in the nineteenth. And only by a similar level of superhuman inspiration could a scientist of the nineteenth century have foreseen the exploitation of nuclear energy that occurred in the twentieth century. But tell me, tell me, who is trying?"

"The usual suspect" I say, "good old NASA".

And it is true: NASA has studied potential missions based on ships powered by fusion, with projects that have already gone through several generations. One model vehicle (only conceptual, unfortunately) is the Discovery II, based on the magnetic confinement of plasma. Using about 800 tons of hydrogen as propellant, and more than ten tons of deuterium as fuel for the fusion reactor, the vehicle could reach speeds of an order of magnitude faster than those of a spacecraft powered by the fission of americium-242. Discovery II could reach 400 km/s, with a payload of 150–200 tons, including the human crew. With such a ship one might think it possible to reach Jupiter in 4 months (one-way). With it could become suddenly possible for manned missions to explore the outer Solar System. However, the features and technical aspects of this engine and the rest of the ship are not known in detail. Nevertheless, the principle would be that the thrust be provided by hot plasma released through a special nozzle, using hydrogen as a propellant.

Moreover, for fusion technology to be applied efficiently to space transport, we need to imagine other solutions, based on different physics to achieve fusion. One, for example, which is called "inertial confinement" to distinguish it from the magnetic field containment, is to implode microspheres of deuterium or tritium by targeting them with a high-power laser. The theory is (or perhaps we should say "was") based on the principle that, if we can manage to concentrate enough energy into a small volume (a few millimetres) by using a laser, we should be able to create the conditions for that magical ignition, i.e. the ignition of the fusion reaction. Today, on Earth at least, we no longer believe so much in utilizing this principle. But it is certain that if it could be applied in a spaceship, the speeds obtainable would still be respectable: more than 150 km/s.

The two possible methods of getting fusion under control, namely magnetic confinement or inertial confinement, appear to be a bit like the contrast between the nineteenth-century steam engine and the internal combustion engine. In magnetic confinement, energy is extracted in the form of heat, which is then used however one wishes—such as in the boiler of a locomotive or a ship. In the inertial confinement system, there occurs instead a rapid series of small explosions (in fact small H-bombs) which directly produce the energy required, as in the petrol or diesel engine.

It is a comparison that has its limits, but it shows in a historical sense how troubled the development of physics and engineering for the production and exploitation of energy for propulsion has been. And not only that, for decades, sail and steam power had coexisted, battling each other to rule the seas; over a similar period, steam engines, internal combustion engines and then electric power had fought for prominence in power supply in railways and in factories. But indeed, there is nothing wrong with having two parallel methods of harnessing fusion.

One method of propulsion by fission–fusion, which may be a little dramatic, is nevertheless interesting. The British-born physicist Freeman Dyson, who was always an original thinker, developed it in the 1960s. It is, quite simply, a spaceship that carries on board a number of H-bombs (with at least as many A-bombs as triggers, of course). At regular intervals, the ship shoots out a bomb from the stern (where the ship is protected by special shock-absorbing armour), where it explodes and the ship accelerates by absorbing some of the energy of the explosion.

"*Ach !!*" says von Braun "I understand, you are talking about that crazy Englishman, under suspicion of communism … I remember him well. But this was not a bad idea of his. Doing the calculations, one can see that his ship, by dint of a series of bangs and jerks, would reach the respectable speed of 10,000 km/s within ten days, which is not bad, is it?"

"Yes, Wernher, but the idea was quickly abandoned when, fortunately, in 1961, the Nuclear Test Ban Treaty was approved prohibiting nuclear tests both on Earth and in space …"

Today, what seems even more interesting is the method called Magnetic Target Fusion (MTF). It is a two-stage process, rather like the combination of the two methods of confinement to achieve fusion. The fuel for departure is confined magnetically while being heated up to become plasma. At this point, inertial confinement occurs, which dramatically increases the density of the material being heated up to its ignition point. In short, this again resembles a continuous barrage of small and very efficient H-bombs. Still, it is a good idea, and once again NASA has worked on it, to the point of coming up with a possible MTF vehicle project.

It is a project that would have many advantages: an MTF vehicle would be smaller than any other fusion-powered spacecraft, due to the efficiency of the double process, and for the same reason it would use fuel more efficiently, in other words it would consume less. Moreover, MTF vehicles could attain higher speeds than other fusion vehicles: up to more than 700 km/s. But there is more: by analysing the MTF process, it turns out that, in reality, only 20 % of the energy produced would be transformed into kinetic energy. A substantial fraction, maybe as much as 70 %, would be lost to space as useless electromagnetic radiation, especially as X-rays. If such energy could be recovered and somehow harnessed and exploited, for example by running a laser propulsion auxiliary engine from it, calculations indicate that terminal speeds of the order of between one-hundredth and one-tenth of the speed of light may be obtained. It seems incredible but it is true and based on sound physics, although entirely theoretical for now.

Imagine, then, having a spaceship, assembled in a low-Earth orbit or even better in a spaceport at the inner Lagrangian point between the Earth and the Moon. Imagine it ready to go and capable of speeds of, say, a few thousand km/s, or one-hundredth of c, the speed of light. With a payload of, say, 200 tons, including a crew of a few people, we would be able to mount a real exploration beyond Mars:

to begin with, in the asteroid belt. Indeed, let us suppose that the ship and its crew have already been on a test mission, perhaps of a couple of months, perhaps to an asteroid approaching the inner regions of the Solar System. We would, at that point, have already refined the techniques of interplanetary navigation and, above all, of the most difficult technique: the approach manoeuver to an object in low gravity such as an asteroid, which cannot be "landed" on in the same way as a planet. And we will also have already learned how astronaut–geologists, armed with their little ice-axes, might explore the very dangerous and unstable surface of a small celestial body that no one has ever seen before; and there it is, the same as it has always been since its formation almost five billion years ago. We would be able to redo, with an intelligent and flexible crew, what the ESA Rosetta mission did on a comet for the first time in 2014. Imagine having astronauts exploring first hand the surface of a comet… and all this just for training.

For a real mission in the asteroid belt, we should be guided by the classical methods of exploration of a century ago, like those of Amundsen and Scott going into the unknown to the South Pole. They knew very well that they could not carry enough food and fuel with them to last the whole trip, so in the previous months, they had quietly built along their way a series of food and fuel depots, to be used primarily on their return journey.

Even with a MTF-powered spaceship, when heading to the asteroid belt, we would not be able to carry everything needed for a return trip. It would be important to launch beforehand reliable unmanned cargo ships, carrying large fuel reserves (hundreds of tons of liquid hydrogen, for example), which should be able to position themselves, perhaps, in a Martian orbit, or at some other quite stable and safe point along the way. We could send the reserves long beforehand, even a few years, using conventional chemical propulsion for their transportation, as we did for the Mars exploration.

With a mission lasting about a year, it would be possible to make a first visit to the asteroid belt between Mars and Jupiter, which is clearly a dangerous place, but fascinating nevertheless. Asteroid 16 Psyche, for example, is a huge object of 200 km diameter, with average density greater than six, apparently made entirely out of iron-group metal and heavier elements. It could be the remnant of the core of a small, never-formed planet … an interesting object, well worth a visit. All the more so, since meanwhile we would have developed certain methods of analysing, at close range, the surface composition and, where possible, the interior of those incredible objects that are the pure and primordial asteroids. We know that this cannot be done easily with unmanned missions. Because here, the courage, intuition and impetuous enthusiasm that are fundamental aspects of the human explorer would be essential. To solve problems that continually offer multiple choices and large numbers of variables to decide between, perhaps rapidly, such as which objects might potentially be both the most interesting to investigate and the safest to reach, the good old human brain is hard to beat.

Let us remember that we are not seeking an impossible asteroid made of pure gold (although …), but we are rather looking for more interesting objects from which we might, first of all, learn about the formation of planets and understand the origin of the Solar System. Then, of course, we could also try to find those asteroids that perhaps have a high content of elements that are particularly rare (e.g. the so-called rare earth elements) but are extremely useful on our planet. By studying their nuclear and atomic structures and from their position in the periodic table, we try to understand, phenomenologically, why they are so rare; so far, the real reason for their rarity is only known to the stars in which they were created. We do not have clear ideas on how to find the "right" asteroids, and we are not thinking of a kind of space gold rush, like that of 1849 to California. We should rather think of the public–private enterprises that will be able to plan for the needs of the planet and then design and fund initiatives aimed at developing such innovation as might be necessary. That means, to look above and beyond the profits that might be had from bringing back to Earth an asteroid (if any) containing tons of osmium or rhodium or cerium or whatever.

Once it will have been well tested and refined, the main tool for the manned exploration of Sphere +3 will be the fusion spaceship (whether of an MTF type or another, we will see). And supposing, as seems theoretically possible, that we can take it up to speeds of, for example, 10,000 km/s, then we would be ready to visit the satellites of Jupiter and Saturn with missions of a duration of a few, say two to ten, years. We are talking about very difficult missions, in terms of their duration, their organization (where to put the supply depots etc.), and the extremely hostile radiation environment that exists around the giant planets. Jupiter in particular has such a strong magnetic field that it results in much particle radiation, protons and electrons, being trapped in the magnetic field around the planet. They would make very problematic the survival of any living being there.

We do not want, therefore, to end up too close to Jupiter and its deadly "Van Allen belts", although we would have screened our spaceship using the best passive and active systems that we would have been able to devise. We also do not need to take such risks, because we are interested, above all, in the system of satellites of the giant planet. To date, we have officially catalogued 67 objects (and counting...) in orbit around Jupiter. This is remarkable progress since the first four were discovered by Galileo four centuries ago: in order of distance from Jupiter: Io, Europa, Ganymede and Callisto.

Some of the Jovian satellites, especially the smaller ones, often have strange orbits, so that they may also travel in retrograde motion to the direction of rotation of the planet. These are clearly passing objects that were captured by the planet's strong gravitational field. The other, larger, satellites were probably born from a veritable equatorial disc of protoplanetary matter rotating around Jupiter: a miniature version of the birth of the solar system from the great proto-planetary disc around our Sun.

The four Medici moons (as Galileo called them, since he wanted to make a career for himself in Florence) are the most interesting. Ganymede, the largest, is a true and proper planet; it is bigger than Mercury. Io is too close to Jupiter to be explored: it is in an inferno of radiation and its surface is pockmarked with volcanoes that spew out lava and sulphur; not because the small world has an intrinsically very hot core, but because of the enormous amount of gravitational energy that Jupiter imparts through a tidal effect. Although this may seem hard to do, tidal forces on Io literally melt rocks by friction from their moving against each other. No, Io is a place to take pictures with our telescopes and to study from afar.

With our fusion supership, we navigate our way, instead, to another satellite also discovered by Galileo, Europa, which is about the same size as our Moon. It is a fascinating subject, among other things, for its diversity: a real complete world in miniature. It probably has a ferrous metal core (and therefore, perhaps, a magnetic field), over which is a thick layer of rock, and finally, the last one hundred (or more) kilometres up to the surface are made of water. Yes, water, like that of our oceans. Indeed, according to the latest calculations, the total mass of water present in the oceans of Europa is about twice that of all the water in the Earth's oceans. Not bad for a small planet a little smaller than our Moon.

Of course, water on Europa also exists as ice at the surface, where the external temperature is about −150 °C. No one knows how thick the layer of ice might be: it seems it could go from hundreds of metres to several kilometres, but certainly, underneath, the water is liquid. It is heated by the gravitational forces of Jupiter and perhaps by volcanic activity from below. In any case, there is no doubt that there is liquid water there. It may even be salty, it seems. And with another interesting feature: within the liquid ocean that envelops the whole of Europa, there could be dissolved more or less the same amount of oxygen as there is in our seas. The physical mechanism that gives rise to the possible presence of oxygen in Europa's water is, at least in part, the dissociation of water molecules in the surface ice, due to its bombardment by cosmic rays. Some of the oxygen molecules that are so created, before being dispersed into space, are able to spread into the water below through cracks in the ice caused by tidal motions. But who knows how these things really are on Europa?

The fact is that in the case of Europa, we have a great opportunity for studying the possible existence of extra-terrestrial life. The conditions for life, namely the presence of liquid water, salts, dissolved oxygen and heat, are certainly present on Europa *today*. They were not present, perhaps, a billion years ago, but they are today, and in very similar forms to those found at the bottom of our oceans, in total darkness near the volcanic fumaroles. It would be enough, perhaps, to pierce the ice, as fishermen do on the frozen lakes on Earth, even if the ice on Europa is perhaps a bit thicker and then let down, through the hole, not a fishing line but a camera with lights... and who knows what we might discover?

All of this, of course, providing a landing can be made on the ice, which is very smooth on Europa.

"What do you have to say about it, Wernher, does it remind you of something?" I ask the great von Braun, who has been listening intently to these updates on planetology, literally opening new worlds.

"Of course, I remember my landing approach with skis on the surface of Mars, arriving at the poles in the winter, although there it is a little chilly. Moreover, you can build a base easily and quickly with the ice, on the ice, if it is thick enough" he replies.

It is already getting old Wernher's enthusiasm going again, the idea of building a base, at least a semi-permanent base, where everything necessary for exploration needs to be prepared on site. All with the added complication of needing to shield against the high level of radiation at the surface, an ever present obstacle because of Jupiter being so very close.

As in the case of Mars, we will have had to have sent beforehand a cargo vessel carrying on board reserves of fuel, the support materials for the shields, generators for power and heat and much more. Above all, we would need a high-power corer, probably nuclear powered, to drill down perhaps miles and miles through the ice, and then, there are the cameras, cables and the rest. And we would have to make sure that everything was well sterilized, unless we want to export to Europa some of our earthly forms of life.

Then, we come to our explorers aboard the faithful MTF fusion driven space-ship, which with its speed is allowing us to visit, study, understand and exploit (well, explore...) a huge part of the Solar System in the very short time of few years. Having arrived on Europa, and having made the hole in the ice, perhaps, we will pull out a "European" fish of a strange aspect. If that were to happen, we could confidently say that yes, that fish will have cost us more than an earthly bass bought at the market; but in order to have been able to see it, an immense road of tech-nological innovation will have been opened to get there, and a great leap forward will have been made in our knowledge.

With the MTF, we will not stop at the Jovian system. We have next to explore Saturn and its satellite Titan, second only to Jupiter's Ganymede. We already know something about Titan after the European Huygens probe landed there in 2005. In fact, it would be fun to go to find the probe, if it has not sunk in the semi-solid methane mud. Titan would be another fascinating place to visit, with its snow falling as flakes of solid methane into lakes of liquid ethane, with Saturn, the lord of the rings, dominating the sky.

Then, far beyond Saturn, a trip to Neptune, which even with MTF, would take more than a decade to get there and back. There we would find the mysterious Triton, the largest satellite of its type, the retrograde ones that run backwards with respect to their planet. It is neatly split in two, just like an ice cream of vanilla and chocolate, and nobody knows why. It may be like Pluto, an object that has come in from the outer parts of the Solar System and was probably captured by the respectable mass of Neptune.

There is a whole bestiary, an immense variety of bodies, major and minor in the Solar System beyond Mars, including a huge population of comets, like the one visited by ESA's Rosetta in 2014. There is a lot yet to be learned from the comets that have come from outside the Solar System: they are probably still pristine, with all their water content and organic molecules, even complex ones, churning there since the origin of the Solar System, about five billion years ago. They must be fascinating to explore and analyse closely.

"Dear Wernher, I see from afar the boundary of the Solar System... we salute you here, and thank you for guiding us and accompanying us to Mars and beyond; in short, for letting us explore Sphere +3 of the Solar System, with its incredible variety of small and big planets. We shall now put our faithful MTF fusion ship in the garage: we are already looking further afield."

"*Ach, nein,*" he says, "do not put it in the garage, or at least do not throw it away. Remember that you need the bicycle even if you have the car, and the car you if you have the plane. So, once developed and gloriously used to arrive at the edge of the Solar System, you can use it again. With it, going to the Moon will be like taking the lift; and to Mars like getting on a tram or a bus. Indeed, I already predict corporate outings for skiing on the slopes of Mount Olympus, where there are the longest sky slopes of the Solar System, and with a gentle gravity that allows you to make lovely, lazy turns…"

A great dreamer old Wernher is, but he does have a point about the possibilities of exploiting Mount Olympus as ski resort.

Chapter 9
The Sphere +4 (or the Fixed Stars)

Homo sapiens Among the Nearby Stars

Having reached the borders of the solar system, there is bound to be a desire to lean on the fence, even if there is no one out there, and to look outwards, and, while having a look around, to wonder what might lay beyond. After all, this is the first time we have arrived at this spot.

But, what's this? I see a human figure leaning against the non-existent-fence!

Wrapped in a strange cloak, staring into the void, drumming his fingers rhythmically (seems rather strange). I seem to recognize a childhood friend. Next to him, leaning against the fence (that still is not there), there is a classic *telum*, one of the fearsome spears used by the Roman legionaries. I am not sure how he got this far; I begin to suspect that he came on foot and is trying to work out if he is at the edge of the Universe, while in the meantime, he is thinking in hexameters. But yes, it is him, it is Titus Lucretius Caro, dear old Lucretius, or Titus to his friends; he who wrote *De Rerum Natura* and put it all in verse, even his shopping list! This is why he counts feet (i.e. the stresses) by drumming his fingers … Let me try something:

Salve multum, Tite Lucreti, poeta preclare! Quid agis?
Recte, ut nunc est. Quis tu?
Joannes Fabricius, stellarum vestigator, qui te interrogare velim.

He looks at me and … does not seem real. He immediately sets off in a philosophical explanation of his presence there. Since the time he wrote his greatest poem, beautifully entitled *De rerum natura* (which today we would translate as *The Theory of Everything*), Titus has travelled (always on foot) in search of the edge of the Universe, with his precious spear at his side.

If the universe is finite, Titus tells me, sooner or later I will arrive at its border. The great poet–philosopher then asks me: if I were to throw this spear outwards from there, where will it go? Will it hit something and stop, or will it go somewhere, beyond the border? *Tertium non datur*, he says, there is no other possibility. Well, if the spear were to hit an obstacle and find itself stuck in something, or if it continued in flight, then the point from which it was thrown was not the last point it

---

Warning: reading this chapter may be harmful to mental health.

© Springer International Publishing Switzerland 2015
G.F. Bignami, *The Mystery of the Seven Spheres*,
DOI 10.1007/978-3-319-17004-6_9

should have been thrown from, and therefore, you have not thrown it from the edge of the universe.

In short, urges Titus, what happens to my *telum*? *"Quid telo denique fiet?"* One will never find a limit, because the possibility of forward flight will be renewed each time. With an elegant mathematical logic, of incredible modernity, it is concluded that the universe must be infinite. Thus, it is proven; indeed, *Q.E.D.: quod erat demonstrandum*. One of the clearest pieces of cosmology in the history of human thought, written, moreover, in the most beautiful of hexameters.

"Gratias ago tibi" I search for an answer, and, in exchange for his lesson, I offer him a ride in my starship.

While he is getting himself aboard my flaming new *"Per aspera ad astra"*, *toga*, *telum* and all, I explain him the theory of the expansion through the spheres of *Homo sapiens*, the cosmopolitan invasive species in love with exploration, and even of the subspecies *sapiens planetarius*. I talk of the Anthropocene as a dangerous new era, in which not even the Olympians could stop most *sapiens* ... these things have always happened even in his own day! But Titus, with his solid philosophical culture, understands everything right away, including the history of the spheres.

I make him put his *telum* down (perhaps it might be of use later; for who knows where we might arrive, but for now, I would not want anyone to get hurt), and I sit him in front of a porthole. I think he is one of the greatest poets humanity has ever produced, and maybe he might like to write something, inspired by the starry sights. I give him a notebook and a ballpoint pen (both things surprise him a lot, but he immediately understands how to use them) in case he might want to write down any hexameters that may come to him, and we leave. Meanwhile, he tells me a strange story, which I honestly understand only partly because he uses a very special *consecutio temporum*, or sequence of tenses, which moreover abounds in ablative absolutes. In short, it seems that while he stood there leaning for ages and ages against the fence-that-was-not-there, he became, quite understandably, a bit bored, but observed a bizarre phenomenon.

"Quater" he tells me, "four times, a few decades ago, I saw strange artifacts flying by. Strange things they were, that passed close by me travelling faster than a hare being chased by dogs on a foggy morning in October (he could not help being a little imaginative). They were going straight on ad infinitum, into the darkness between the stars. They bore strange insignia, not the Roman eagles nor fasces. As they flew by, I saw on them a peculiar blue label, with white and red squiggles."

After a while, I realize what he has seen ...

"Dear poet, now I understand: what you saw flying by they were the two Pioneer and the two Voyager probes, the only man-made objects to have left the solar system. The insignia you saw is NASA's emblem, which, unlike your legions, does not include fasces nor eagles... But NASA is a very a powerful organisation in America ..."

Oh yes, I realize I will also have to tell him about the discovery of America, and a few other details. Above all, before establishing a flight plan, I will have to explain to him what the state of the art is in the fields of astrophysics and space technology. While I do that, let's take stock of the situation.

In the context of our spheres, we define Sphere +4 as the one extending outwards from the boundary of our Solar System and going out so far as to encompass a number of the nearest stars in our galactic neighbourhood. Within twenty light years from the Sun, there are nearly a hundred stars, which are easy to study because they are nearby. Let us thus define Sphere +4 as that of "nearby stars", having a maximum radius of twenty light years.

Now, we still do not know how to get there, but at least we know where to go. Moreover, we already know today (and tomorrow we will know even better) a reason for going there. It is very simple: we are finding more and more planets around stars, even those nearby. Perhaps E. T. lives on one of these and we will want to go and meet him!

In this sphere, then, there now seems to be a concrete chance that there are worlds other than the Earth to be explored, and which are perhaps comparable with it. It is an argument even stronger than that which drove Marco Polo, Columbus, Magellan, Cook or Amundsen to explore new, unknown and possibly hostile continents and oceans on Earth, but where at least they know they would have air to breathe. The topic has become even stronger recently, with the discovery of a terrestrial planet around the star Alpha Centauri B, one of the closest to the Earth, at just over 4 light years from the Solar System. And we know that this is a rocky planet, just like our Earth, not a gas giant like Jupiter or Saturn. Rocky, and with a mass which is not much more than that of the Earth. It would therefore seem to be a very interesting planet, except for one serious flaw: it is very close to its star, around which it orbits in just a few days. The temperature on its surface, therefore, is thousands of degrees, and it is certainly not a suitable place for life.

But around the same star, there may well be other rocky planets. We may just have found a kind of Mercury from another Solar System. A little farther out, at the right distance from Alpha Centauri B for water to be maintained in its liquid state, and which therefore could potentially support life, there could be another Earth, not yet discovered. We should remember that the method used by the two Swiss astronomers to find the planet Alpha Centauri B (using European telescopes in Chile) favours the discovery of planets close to their parent star. Those more distant, and therefore more difficult to be detected, may nevertheless exist, but still await discovery. In fact, the whole Alpha Centauri system, which is made of three stars held together by a complex gravitational orbit, is a new discovery for the emerging science of planetary astronomy. Just because it is a multistar system, it seemed to preclude the formation of planets, at least according to calculations made so far.

We know that planets form around a star (like those of the Solar System around the Sun) from the same disc of interstellar matter from which a star collapses from its parent cloud. What remains of the proto-stellar material, after the formation of the star, is concentrated in a disc that rotates around the equatorial plane of the star. At a rate that is not yet well known, but which may be quite high, there form from this disc at first small aggregations of matter, which then grow bigger and bigger, like lumps in porridge. As they grow, the lumps of matter become planets, of varying compositions and sizes.

The whole process of the formation of a planetary system—that is the collapse of an interstellar cloud into its protostar and proto-planetary disc, to the creation of a planet like Earth—only seemed to be possible around a single star. Once formed, albeit as a fairly violent and catastrophic event, it would create around itself the right environment for the formation of planets, just as occurred with our Sun.

However, a double or triple system of gravitationally bound stars seemed to introduce such severe gravitational perturbations to the delicate process of planetary formation, that any proto-planetary disc, it was thought, would be destroyed or scattered by the catastrophic passage of the other stars in the system. The discovery of a planet orbiting Alpha Centauri B has provided definite evidence that since one star in this well-known triple system has a planet, then it is not only single stars that can form planets.

If, as we suspect, this planet is part of a true planetary system, then our hopes of finding a planet similar to Earth, placed at the right distance from the right sort of star, are very much increased. In short, we had better get ready to go and explore some nearby stars, especially, since, in our neighbourhood, the numbers of stars with planets seem to be sprouting like mushrooms.

Beyond Alpha Centauri, there are also Epsilon Eridani and Tau Ceti, which are single stars, apparently with planetary systems, and all well within the arbitrary limit of 20 light years of Sphere +4. Indeed, Epsilon Eridani, at least, has a rocky planet, with a mass not too different from that of the Earth, orbiting at the right distance. Liquid water may even exist on this rocky neighbour. But discoveries such as these are only our very first, there are certainly more interesting objects orbiting around nearby stars waiting for us to discover them. To put things into historical perspective, we should remember that the first attempts to search for intelligent extraterrestrial life were made half a century ago by Francis Drake with the Ozma Project, the precursor of the more ambitious SETI project. Just half a century ago, Drake began with two nearby stars with something "promising" about them: they were Epsilon Eridani and Tau Ceti.

The distances we have to travel to reach Sphere +4 are, once again, larger by a factor of a thousand than the distance we had to cover to reach the fence-that-does-not-exist around the Solar System, where we picked up our passenger (who now, on the bridge of *Per aspera ad astra*, is complaining about the lack of a thermal spa with a proper *caldarium*). Let us say that one has to cross a few light years, instead of one or two light days, which are the maximum outer dimension of the Solar System. This is much the same change, in distance order of magnitude, that separates the sphere of the Moon (+1) from that of a low Earth orbit, or, to go in the other direction, from the Moon to Mars. Do not panic: we already know the answer —one has to be able to travel ever faster in order to be able to follow this intriguing harmony of the spheres.

And in order to go faster, we already know that the recipe is the same: a better fuel, which is to say more "efficient", i.e. with a better energy content per unit mass. After the oxidation reaction of solid fuels (such as gunpowder, which is what we might find in Chinese fireworks), we move up to the oxidation of hydrogen to form the water molecule, which is the best that chemical power can give us. The process that recreates the water molecule generates such a lot of energy because, conversely, it is so difficult to break the water molecule apart, once it was formed. After obtaining a speed of a few km/s with this "water" rocket, which was enough to take us to the Moon, there followed the best that nuclear fission could offer with the nuclei of plutonium or americium to take us up to a few tens of km/s, a speed that might take us to Mars. Then, there was the fusion of nuclei of deuterium, tritium and helium, which accelerated us up to a few hundred km/s, or even a thousand, which might take us around inside the Solar System. Now it is the time to get serious, or at least to dare to do even more.

To reach one of these exoplanets, and thus to become subspecies *sidereus*, *Homo sapiens* will need to achieve speeds in the order of many thousands of km/s, which is a significant fraction of the speed of light: the largest possible fraction of it we can manage. Even so, the *Per aspera* will still take a long time just to get to Alpha Centauri, or Epsilon Eridani and Tau Ceti (and perhaps come back?). But at least, we have a project on which we can think (or dream) of embarking, and a journey that would perhaps have to rely on two generations on board.

We are deliberately vague about the timing, speed and the exact target of a mission of exploration–emigration for those Pilgrim Fathers who will land on a planet of a nearby star, instead of landing on Plymouth Rock. It is only conceptually that we see how we might drive the *Per aspera,* of a mass of a few thousand tons, up to a speed which would need to be a significant fraction of the speed of light. The conceptual answer is by the annihilation of matter and antimatter.

The idea of using antimatter as a source of energy for a spacecraft has been around for a long time. And it is not surprising: the annihilation of a particle of what we (incurably anthropocentric) call "matter" (i.e. electrons and protons) and its counterpart made of antimatter (positrons, antiprotons …) has the highest known ratio between the energy produced and the masses involved. "Annihilation" itself means, in this case, that when matter and antimatter meet, both disappear and its equivalent is released as energy, in accordance with Einstein's equation: $E = mc^2$. The multiplication of mass (m) by the square of the speed of light, which is a large number and is represented by (c), means that it is immediately understood that even a small amount of mass will give rise to a lot of energy.

Of course, there is a problem; in fact, there are several. First, the energy of which the Einstein equation speaks of is emitted mainly in the form of electromagnetic radiation, that is, photons, with a wavelength related to the specific process in question (i.e. whether it is an electron–positron annihilation, or a proton–antiproton). Moreover, the energy is emitted as gamma ray photons, which are difficult to exploit, and a number of unstable of charged particles (e.g. mesons) are also created along with neutral particles, which are also difficult to exploit. Furthermore, a lot of neutrinos are also produced: they carry away energy, but are useless. To take advantage of the annihilation energy very efficiently, one thus needs to consider carefully what to do with all the energy that is bound up preferably in gamma rays and particles, and this is not easy. With neutrinos, however, it is best just to let them go.

A second problem, which may be even worse, is the difficulty of producing antimatter. Unfortunately, antimatter does not exist, even underground, in the same way that oil does from which we can make gasoline, or uranium for fission; or in the same way that the water molecule contains hydrogen, which can be used for deuterium fusion. To produce antimatter, we have to use a lot of energy. On our Earth, made up of "normal" matter, the only way for antimatter to be "born" is by using the opposite process to annihilation: the creation (in equal parts) of matter and antimatter from energy. The necessary conditions are very difficult to achieve. We need to accelerate particles (matter) at very high energies, then smash them together (or to put it more elegantly, arrange for them to interact) and then exploit the products of the interaction, among which there will be our precious antimatter. It can be done, and we know how to do it; but at present, it is a slow, unproductive and expensive process.

A third problem, no less serious than those already mentioned, is what to do with antimatter, once created. This, as we have already said, is not a novel by Dan Brown, and we cannot afford (too many) flights of imagination and invent a kind of magnetic thermos flask or whatever to be filled with antimatter. As is now clear, if put into contact with any container of "earthly" material, then: *puff*!!—the anti-matter (and a bit of matter) is no more. So we have to invent some new sort of apparatus based on vacuums, the cold, and on magnetic fields to hold it almost still, with nothing around it and somehow suspended there or "levitating"... it is for-tunate that we speak of charged antimatter, typically antiprotons or positrons.

In short then, to create, store and use antimatter as a potential form of energy is very difficult and therefore expensive. Moreover, antimatter, although so energy efficient, is not easy to use; this is immediately obvious, since we have not yet made an antimatter bomb to get ahead of the "others" (whoever they may be, but we must always invent "them" and make better bombs than theirs). It would be logical to expect an antimatter bomb: *Homo sapiens* has always immediately made bombs after the discovery of anything suitable that will release energy: the chemical reactions in gunpowder or dynamite; the physics of the chain reaction in uranium for the atomic bomb; or in the hydrogen–deuterium reaction of the H-bomb. From all energy sources, as soon as they can, our *sapiens* go: one, two, three: BOOM!!, with more or less spectacular results, which continuously grow ever more deadly. But do not despair (I turn to any readers who may be warmongers): even at the time when the first atomic bomb was planned to be built, it seemed impossible for the responsible research unit to produce enough uranium 235, of bomb-grade purity, to actually make a bomb. It was extracted from the mass of natural uranium, which is mainly composed of the isotope 238, and of which only about 0.7 % is the isotope 235. The extraction method seemed to proceed literally atom by atom, and it seemed that they would never acquire the kilograms of U235 atoms needed for a critical mass ... but in about two years, they succeeded perfectly well, and it was only their first attempt.

So, even the massive production of antimatter, if it became a real interest for military purposes, could receive a major breakthrough. Should we fabricate a super-bomb in order to build a spaceship, then? No thanks, let us try to learn from our past mistakes. Let us study how to make antimatter as a potentially great source of energy and then perhaps move on to power interstellar exploration, but let us stop there.

Let us see how we might think of using this absolutely magical (but real) fuel from which we can expect huge efficiencies. The annihilation of every gram of antimatter produces a billion times more energy than the oxidation of a gram of gasoline and a thousand times more than the fission of a gram of uranium in a reactor. It just seems to be the right fuel to travel great distances by going very fast.

There are different ways to exploit the process of annihilation. The easiest, if may be a little trivial, is a kind of steam locomotive: the energy of annihilation is used to heat any fluid, and then, this hot fluid can be used to propel the spacecraft. The speed obtainable by this method (less than 100 km/s) is comparable to the Rubbia fission engine: i.e. excellent, but not interstellar, and in any case, the Rubbia engine is infinitely easier and much less expensive to be built. Unfortunately, then, this method of exploiting annihilation may not be of interest to us if we want to visit E. T.

Another system uses antiprotons to "catalyse", in some way, nuclear fission reactions, by firing antiprotons into uranium. The fission reactions are in turn (and here emerges the military tradition in these studies!) used to trigger deuterium–tritium fusion, and once fusion is initiated, the spaceship is propelled … though again only at a little more than 100 km/s. Alternatively, if the first "atomic bomb" stage is skipped and the antiprotons are used to trigger fusion, then this method allows much higher speeds (1,000 km/s) than were attained by previous, just-fusion engines; but since this is only a complicated way to obtain fusion, it may not be worth it.

And what about fuel consumption? It is interesting to look at the latter two cases. In the first case (trigger fission–fusion), precise calculations have been made: carrying a payload of 100 tons to Jupiter and back within a year would use 10 μg (millionths of a gram) of antimatter. In the second case of fusion ignition only, one might imagine a trip to the Oort cloud of long-period comets at the edge of the Solar System, and back, in 50 years (!) using just 100 μg of precious fuel. Not bad, but we must keep in mind the problem of fuel consumption: the antimatter is by far the "thing" (or anti-thing?) that is the most expensive to be created on Earth or elsewhere.

We come now to the most promising method yet for the exploitation of antimatter in space propulsion. It is also the most studied one and has been planned in detail for over twenty years. It is called "beamed core" or irradiated core. The secret here is that all intermediate steps are jumped over in order to directly exploit the products of annihilation. Or, rather, we exploit the easier ones, that is, the charged particles. Because these are created at speeds close to that of light, the whole secret depends on trying to eject them from the nozzle of a rocket exhaust at these high speeds, and well collimated if possible. The key lies in collimation, utilizing a magnetic nozzle, which until recently still seemed in need of being invented.

However, through the brilliant work of physicists and engineers at Kent State University (Ohio) based on computer simulations, a realistic and efficient magnetic nozzle is now being designed: at 4 mt. long and one and a half metres wide, it seems purpose-built for a spaceship. With magnetic field values that are easily obtainable today (less than 200 T, which is a trifling field strength for lovers of the genre), the proposed configuration reaches extremely interesting efficiencies of almost 40 %. In short, the whole nozzle, which is the essential part of the engine, could already be built today or tomorrow at most.

But be careful: with this project, the particles would be expelled at the staggering speed of 0.7 c, and 0.7 of the speed of light means 210,000 km/s!

No panic: this does not mean the same speed for the spacecraft. There are many factors to be considered, such as the mass to be transported in the form of both the payload and fuel, and then, there are those confounded losses of efficiency in collecting the energy of annihilation, losses which are associated with those neutral particles that do not care about the magnetic field and with their speed and their mass carry energy away without doing anything. And then, there are those bothersome neutrinos, which have hardly any mass, but there are a lot of them. There is nothing that can be done with them anyway; we just have to write off almost half of the energy budget. For neutral particles with mass, however (e.g. neutrons), we hope to be able to invent a few absorbing screens that will not let them escape and so will allow us to recover a bit of energy.

In the end, by doing realistic calculations, it is possible to think of reaching relativistic speeds, say a third of c, or 100,000 km/s. E. T., here we come!!! Oh yes, because if these speeds become possible, then access to Sphere +4, less than 20 light years away, becomes conceivable, although it would have to be done on a time scale of a few decades. As these speeds become obtainable, we might begin to see the light at the end of the tunnel.

Unfortunately, the antimatter "engine" is just one of the many problems to be solved. Antimatter production, i.e. the fuel needed, is the real obstacle, and seemingly insurmountable. However, if we think about the history of rocketry and the way in which apparently insurmountable obstacles were overcome, we find the following encouraging precedent.

About a century ago, some visionaries imagined the use of a liquid propellant for rockets as being a big step forward compared to the solid fuels that had already been invented by the Chinese many centuries earlier. Indeed, it was clear that the ideal chemical propellant would have been liquid oxygen and liquid hydrogen, because of the great potential energy of the water molecule. But while there were no great difficulties in getting the oxygen into liquid form, it was a very different case for liquid hydrogen. At that time, it seemed a miracle that just a few years earlier, in 1898, Sir James Dewar had for the first time liquefied a very small amount of hydrogen. But to produce enough to fuel a rocket, to carry it around in liquid form at −253 °C and to have to manage it through pipes, valves, etc., in order to run a motor seemed a highly expensive and highly dangerous folly. In reality, it was not much different from the problem of the mass production and management of antimatter today. However, half a century after Dewar, the everyday use of liquid hydrogen became a reality. Indeed, the epic flights of the Saturn V to the Moon or the Shuttle to the ISS are today already history, with each flight having been based on tons and tons of liquid hydrogen.

So where do we stand now, first of all, with the production of antimatter? Rather behind, at the moment. We are still at the point at which most of the antimatter particles that we produce are either antiprotons (negative protons) or positrons (positive electrons) and are counted one by one, as were the first nuclei of uranium 235, when the method of separating them from uranium 238 was first invented. Furthermore, the antiparticles are only produced in large accelerators, like those at Fermilab in Chicago and at CERN in Geneva.

At CERN, we are now managing to make 10^{15} antiprotons per year. At this rate, to make 1 g of antimatter, it would take a hundred million years or so. We are not there yet! At CERN, though, they are a bit better with positrons, which although energetic per se, owing to their small masses return two thousand times less energy than antiprotons. At Fermilab, where they are more interested in the specific problem and have therefore developed dedicated machines, they are a bit better at it, about one order of magnitude better. In short, if we really aimed to have several micrograms (and that would not be a small amount to begin with), then in theory, Fermilab could now produce it within a few decades.

Do not forget that in the half-century that separates us from their discovery, the intensity of antiproton beams has increased more than ten thousand-fold, and this is well-documented history of science, not fantasy. And this increase has not been linear, but was greatly accelerated. Who knows what our physicists will be capable of achieving during the next half-century, in terms of production of antiprotons, if they put their minds to it.

Then, there is the no less thorny problem, already mentioned, concerning the conservation of antiprotons, once they have been created. Fermilab and CERN manage to do something special with "Penning traps" (named after the inventor), but quantitatively, we are still better at creating them than storing them. However, even NASA is quietly working on the problem of conserving antimatter, so far with modest but encouraging results. It seems that we can store a billion antiprotons for a year, which is, unfortunately, a long way from what an interstellar spaceship would actually need, but it is a start.

Not to mention the costs: while it is possible to imagine production efficiency increasing in the coming years, the cost will always be high, close to something like 25 billion \$ per gram. Although it seems incredible, we should remember that 1 g of antimatter would cost roughly as much as it would cost to fuel the Shuttle a thousand times and would produce the same amount of energy. Viewed in this way, in the end, the problem takes on another aspect: that of producing and storing it as quickly as possible and not to make it known to those who would use it to make a bomb. But I am afraid they already know.

Supposing we were to do it, how would we organize a mission to a nearby star? How would we finance it? As a one-way trip or as a return trip? It is amazing how many people dare to think in terms of a one-way mission. There are those who think that going to another planet would be as it was for the English settlers who emigrated to North America or Australia, when those lands were still empty continents (at least with respect to pale-faced *sapiens*) that were waiting to be filled up with new arrivals.

I believe that it would be more prudent to prepare for a return mission, even if the total duration were to take decades. No, we do not dare delve into the practical, psychological, social and other (religious?) complexities of such an adventure.

We can only say that, following the logic of the expansion through the spheres, *Homo sapiens* will come to the great leap that will turn him into a *sidereus* in a much better prepared state than we are now.

Thanks to advances in astronomy, we will have a much more detailed and accurate knowledge of stellar and planetary systems outside, but close by, our own; we will have in hand the results of space exploration by more robotic, one-way missions; we will have accumulated much more experience on interplanetary flight, in all its numerous interdisciplinary aspects, starting with propulsion. And just with regard to propulsion, basic research in plasma and particle physics, as with all basic research, will eventually put within our reach, or in reach of our engineers, the last two major energy processes of nature that we still lack, namely fusion and annihilation. Today, we know what they are. We do not yet know them well enough to use them, but we will learn, and we will have a great use for them when we do, in space exploration as well as on Earth.

Meanwhile, Lucretius (Titus to his friends) was getting a bit too comfortable aboard the *Per aspera ad astra*. Tired, at last, of writing hexameters, he began to shower me with questions. He understood that we were travelling between the stars (well, of course, just look out the window) and had made up his mind to learn how to operate a spaceship.

"*Tite, dilecte mihi*" I sought to flatter him. "You are a great poet, thinker and philosopher, but in these last two thousand years, mathematics and physics have gone very far. Zero, negative numbers … and now antimatter. I do not think I can explain you how to drive the spaceship".

He said, "*Rem tene, verba sequentur*, said once a prominent lawyer who was a contemporary of mine … if you know what you mean, words will come. Try me, I dare you".

Faced with a direct challenge, I decide to start from elementary algebra and to relate him the parable of the negative piglet (which I invented on the spot). It will be a cultural revolution for him, I know, but I will try.

The Parable of the Negative Piglet

On a bright summer morning, a swineherd was whistling happily, sitting on his cart on his way to the market. He was being pulled slowly but surely by a pair of oxen, and in addition to the swineherd, seated on the box on the wagon, were three beautiful, fat piglets, ready for market. Along the way, he saw beside the road a *viator*. The man had a distinguished air about him and was rather dark skinned, seeming almost Ethiopian.

"*Salve, amice,*" said the fellow, who seemed a nice chap, and added, "*Quo vadis?*" "*Ad forum* … (in a well-known city in the South of Italy), to sell my piglets" replied the swineherd. And the foreigner, in a slightly exotic accent, asked: "*Visne me tibi comitem adjungere?*" Would you not want me to join you, to keep you company? (which was an elegant way to try to hitch a ride).

"Sanequidem, conscende carrum, et apud me asside" but certainly, jump up and sit down here beside me. The fellow proves to be an Indian mathematician, who, for many years, has been studying in Greece at the Academy of Athens. Now, he is in Italy for a conference of philosophers, astronomers and mathematicians (who were, at that time, all indistinguishable from each other, especially to a swineherd...).

"Dic, quaeso, quod est officium matematicorum?" The poor swineherd does not know what mathematicians do, and, by way of conversation, he puts the question to his strange guest. From whom, most kindly, he hears the answer *"Varium et mutibile*, but mostly we do the equations".

"Aequationes?" said the swineherd, a bit confused, while the oxen quietly continued plodding along on the road to the market, which they knew so well. "What are they? ... Give me an example? But an easy one, because your strange accent already confuses me ..."

"Let's see ... So, *latine loquimur*: here's an example of an equation, made with numbers. I'll write it with my finger on the palm of your hand, so you can understand it better". He takes the swineherd's hand and, on its dirty, calloused palm, the Eastern intellectual writes with his own thin brown finger:

$$3 - 4 + 1 = 0.$$

(Note, we are talking about an Indian mathematician, who had already discovered negative numbers and zero ... concepts unknown to you Romans and objectively difficult for a swineherd of that time, and maybe for you too, *dilecte Tite*.)

"I do not understand," said the swineherd, confused and a little humiliated by this stranger.

"Recte, then I'll try with an example. The zero (*nullum?*) would mean, in your case, having an empty cart, without pigs".

And the swineherd: "Even I understand that." And the Indian: "Now imagine that you have to stop for a moment to take a pee, as you say, *mingere*—to urinate, and while you're distracted, not three, but four piglets escape from the wagon...."

Wide-eyed the swineherd utters: *"Portentum!"*

"Certainly, but the meaning of *aequatio* is that if, at that point, you, in desperation, put on the wagon another piglet (which you happen to find on your journey), you will precisely arrive at the market with an empty cart ..."

Inevitably, it all ends badly for the Indian mathematician. He is chased out of the cart by a mouthful of evil words from the swineherd, who does not like to joke about his pigs, whether they are positive or negative, and especially does not at all like this thing called zero, which would mean an empty wagon ...

After the parable, to which he listened carefully, I see with pleasure and with little surprise that Lucretius, a poet and a philosopher and not a swineherd, understood everything. At that point, it is simplicity itself to explain that antimatter is to matter what a negative a pig is to an everyday pig and that ultimately, the wagon is not quite empty, and it is actually full of gamma rays and other particles that pop out. Perhaps this is why, after the annihilation of the pigs, the swineherd's cart sets off again, not behind but in front of the oxen, and at speeds far greater than that of the famous October hare.

Dear Titus enthusiastically tells me

"But then we can really explore other worlds! Tell me, where are you taking me?" I immediately display on the wall a large chart of the heavens, which he regards with fascination. Then I do better: I decide to amaze him with special effects and turn on the giant Digital Globe with which the large control room on the bridge of the *Per aspera* is equipped.

This is taken from a large globe of 2 m in diameter, similar in size to the celestial globe that the Franciscan scientist Vincenzo Coronelli made for the Sun King, Louis XIV at then of the 1600s, but it is actually an innovative spherical digital screen. The result is a bit like the "Palantír of Orthanc" in the *Lord of the Rings* by JRR Tolkien. Thanks to the internal projection, it is geometrically perfect, and with a touch on the keyboard, I can light up on the surface of the Digital Globe all the constellations around the Solar System and also draw the great circle route of the *Per aspera* in three dimensions. He has his hexameters; we have technology with high visual impact. Titus is shocked at first and then gets involved. He becomes absorbed in the image and then looks at me, waiting.

"Here is where we are, Tite: at this point we have a choice. Of the planets closest to us, there is one in orbit around Alpha Centauri, confirmed to be of a terrestrial type and perhaps habitable, but on which we do not know enough, so far; and then it is within a messy star system. Then there's this one, in orbit around Epsilon Eridani, which is also of a terrestrial type and also quite nearby; it may be interesting to try it. Here is a catalogue that NASA has just compiled, containing a list of the potentially "best habitable" rocky planets, ranked by an index called the ESI (Earth Similarity Index), which lists them all according to how much they are similar to our Earth".

The ESI is real and truly takes into account everything, including mass, surface temperature, presence of atmosphere, of a magnetic field, etc. ESI has a value of 1 for the Earth, of course, and decreases as we move further away from terrestrial conditions. Mars, for example, has an ESI of 0.6. But be careful: planets with an ESI greater than 0.7 can host elementary life forms, while those ranked at 0.8 or more could accommodate more highly evolved life forms.

"We select which planets to go to, then, dear Lucretius, according to both their distance and their ESI value. The result is very interesting: it appears that Tau Ceti, which has two planets (Tau Ceti e, the fourth, and Tau Ceti f, the fifth) that have an ESI of 0.77 and 0.71, respectively. Values that allow, perhaps, the existence of mould, bacteria, and at most a few little worms, or similar elementary forms of life, so removed (we think) from our own, but better than nothing. These two exoplanets are 12 light years away, and have masses 5–7 times that of the Earth, so they are perhaps a little too big for our muscles, but with a little training and a few generations of time, we may be able to cope with their gravity. On Tau Ceti e, the average surface temperature is 68 °C, which may be a little too hot, like summer in the Sahara, but not impossible for life. On Tau Ceti f, however, temperatures are around −35 °C, like a winter in Siberia, where life also abounds".

But what appears to be even better on our Digital Globe is in the constellation Libra. There is a star, Gliese 581, named after its discoverer, an avid hunter of stars, just twenty light years from us. Around Gliese 581 turns a real system of many planets, two of which (d and g) have an ESI above the threshold. Indeed, g, with a mass of only 2.6 Earth masses and a surface temperature of +10 °C, has an ESI of 0.92! It seems like a pretty exciting place … Planet d, however, is a bit plumper (6.9 Earth masses) and much cooler: −37 °C, the stuff of Antarctica. However, its ESI is 0.72, which is better than nothing.

Dear reader, everything recounted above to Lucretius (who listened to me open-mouthed, staring into the Digital Globe) is all strictly true, to the best of our current astronomical knowledge. In other words, less than twenty light years away (the radius of Sphere +4), there are at least four planets that hold a high position in the ESI ranking, all of which are above the threshold for the possibility of life. In addition, there are at least two other very interesting planets, in orbit around Alpha Centauri and Epsilon Eridani.

Let us remember that the first extrasolar planet was discovered in 1995, and today, we know of more than two thousand of them, and their number is increasing. Remember above all that just a few years ago, when we were celebrating the 400th anniversary of the discovery of the Medici Planets by Galileo in January 1610, the number of potentially habitable planets (ESI > 0.7) and placed in Sphere +4 was equal to … ZERO. Today, the number of nearby places where E. T. could be living is at least four, and they are destined to grow rapidly in number and as well as in the quality of their environment. It is worth thinking about it and then working hard towards building for real the *Per aspera ad astra*.

Coda (Less Romantic, but We Explain the Mystery)

Stop! If you have opened the book at this page, TURN BACK. Do not read the explanation of the mystery of the seven spheres immediately. It's not worth it; first you must have read the whole story. One is granted access to the coda only after the end of the story, as in the caudate sonnet, which adds a septenary and a couplet to crown the canonical double quatrain and double tercet. Even in music one might expect to hear a coda at the end of a concerto when, after the three canonical movements, the composer's sense of completeness is not quite satisfied.

In our case, much more modestly, we use a coda to explain the mystery of how the future of manned interplanetary and interstellar space exploration might be realised (i.e., how to finance it).

Meanwhile, we shall summarize in a few words and figures the history of exploration, recent and otherwise, by *Homo sapiens*, that cosmopolitan invasive creature like mice, cockroaches, and crows, and who, like those other species, is also an excellent explorer.

SPHERE 0. Most of it has been explored in the last 143,000 years, that is, since African Eve, our common ancestor from whom of all of us, today, have received our mitochondria. The vast majority of the land and sea surface has been recognized, measured, photographed, if not trampled over and sailed across by a *sapiens*. Mankind's investment in the exploration of Sphere 0 has been done very well. But that is the past. Little now remains to be explored.

SPHERE −1. We still know very little, however, about the submerged part of our planet, which covers the majority of its surface. We started quite well and we are making rapid and tremendous progress, although, perhaps, we are driven more by our interest in profit than from the noble fire that burns in the heart of the explorer. But we have not yet begun to collect the billions of tons of gold or uranium dissolved in seawater. Humanity's investment in Sphere −1 has been small, compared to what remains to be explored and the potentially tremendous results that might be gained; perhaps a billion dollars, in the last century. There is still much left to discover and invent.

SPHERE −2. Here we have done really badly. Except for some ridiculous little hole drilled down to a little more than one or two thousandths of the radius of the planet, we have only indirect information. Although a valid and important area for exploration if we wanted to do it, geophysical exploration is still very modest: shall

© Springer International Publishing Switzerland 2015
G.F. Bignami, *The Mystery of the Seven Spheres*,
DOI 10.1007/978-3-319-17004-6

we say a billion dollars have been spent here in the last century. There is everything still to do, starting with the "underground satellite." It is a project that would cost very little (maybe a few billion) with respect to the potential cultural and practical knowledge it would yield.

And now we look at *the pyramid of extra-terrestrial exploration*, by humans as well as robots. The future belongs to the boundless space beyond the Earth. This is where we have so far made the largest investments and obtained the most exciting results from which to continue the adventure.

SPHERE +1. Here the environment is now becoming overcrowded. Thousands of satellites occupy various orbits; and the lowest possible orbits, such as the low altitude polar orbit, are densely populated with satellites almost in sight of each other and at high risk of collision. We seriously urge regulation. From Gagarin to date, at least 500 humans have visited Sphere +1. At least 18 have not returned and we remember them in silence. Even here, little remains for us to explore, but much to be done for private profit and to improve further the quality of life of *Homo sapiens*.

SPHERE +2. In addition to 24 human feet, we have laid on the Moon hundreds of tons of earthly material, while we have brought back to Earth, less than half a ton of lunar rocks in total. Also, near or around the moon hundreds of probes have passed. It's not much, but it's enough to get an idea. The idea was, of course, that sooner or later we would have to go back there. Unfortunately, what with one thing and another, we have never found the time to do it.

SPHERE +3. Men, zero; robotic probes, quite a few. To be precise: to Mars at least 50, beyond Mars 14. We have learned how the Solar System formed, but we do not have the courage to go there in person, not even to start with, just to Mars. Yet we now know how to do so.

SPHERE +4. Men, zero; probes 4. They have not yet arrived at nearby stars, but in short, four man-made objects have made a timid exit from the Solar System. The seventh sphere is still substantially devoid of any trace of humanity.

We are still in the field of exploration of the four extra-terrestrial spheres and in that respect we also need to talk about money. In just over half a century, from Sputnik in 1957 to today, we have spent less than half a trillion dollars (five hundred billion) on space exploration (i.e., the total sum spent on the exploration of Spheres +1, +2, +3, and +4), including the ISS, the Shuttle and the Apollo programmes which, alone, have had the lion's share at more than half of the total space budget. Yet this is a trivial amount if one compares what has been spent on "civil" space with what has been spent globally on military interests. We saw this in sufficient detail in Chap. 6: Current official estimates of military spending are 1.7 *trillion* dollars per year. *Per Year*. It is therefore clear that we must go fishing in the military budget in order to save space exploration, and so give it a greater share, especially for Spheres +3 and +4.

Let's step back, though, and we see in general how a large programme of space exploration and interplanetary and interstellar colonization might be financed. Let's start by saying that, to finance such a major challenge as the human exploration of

Sphere +3 and especially of Sphere +4, it would certainly require the mobilization of the public as well as the private sector.

We would propose that the public sector should take charge of building a spaceport close to the Earth, and running its operations (which would have significant costs over the years). The public sector should also have the responsibility of the research and construction of robotic probes for interstellar and planetary exploration, as well as the research and development of spacecraft for manned exploration, including the methods of propulsion.

Private funding, however, will use both the fundamental and applied results obtained from public research in planetary exploration and propulsion. Private individuals, as we shall see, will be mainly involved in the exploration and colonization of those planets that have been judged to be interesting from astronomical and robotic exploration. The distinction between the role of private and public funding follows the principles of general economics which are now well entrenched. In fact, the implementation of services and projects in order to create a benefit for all should be financed publicly, because the entire audience (i.e., the taxpaying population) will benefit. Naturally, the allocation of public funds, in addition to those for fundamental research, which is the most basic of all public services, and as such is indisputable, will be calibrated according to the public's demand for certain services and projects.

Clearly, in the case of funds needed for an expedition into Sphere +4, the bulk will be allocated when there has been an unequivocal discovery of habitable planets near us. Such a discovery will drive the interest of the general public first to "go and see" and soon after to "try to remain there" and then, who knows. It will therefore be due to there being evidence of an interest among the general public, generated by discoveries made with public funds, that will push private investment toward the final stage of exploration and colonization. It is this phase that, so far, is completely missing from Man's history in space, and is the one that could provide the edge over our current state as *sapiens*, for us to become first *planetarius* and then *sidereus*.

Finding public resources for the first phase—that of research and preliminary exploration—will not be easy, because it still demands large sums of money. The recipe can only be the same as that which has already mentioned: a gradual reduction of the enormous investment in military spending in favour of an increase in all fields of space research and exploration. It is the only way to achieve the necessary advancement without affecting other important areas of public expenditure that every government has to fund, namely education, health, the environment, etc.

In 30 years, the world has spent at least 50 trillion dollars on conventional and nuclear weapons. The total spent on space, in more than half a century from Sputnik to today, is rather less than 2 % of that amount. That may seem unbelievable, but it is true. To begin with, by simply redirecting only 2 % of that currently spent on arms, we could immediately double the investment in space exploration. If such an agreement could be reached at least between the most important States, then the other States would soon follow. It should finally lead us toward a culture of non-violence and in addition there would be a large positive effect in providing an adequate challenge to the same type of industrial complexes that now produce weapons. Indeed, it would provide a formidable impetus to the development of a completely new technology, even better than that which the military already experiences (which, unfortunately, is quite good). Let us try to assess the overall cost (both public and private) of a long-term programme of "interstellar" exploration, i.e., including our landing on the closest planet that is judged to be "habitable" (and we have seen that we have already discovered many within 20 light years).

For a reasonable costing, we shall divide the problem into three phases: (1) the invention and development of new propulsion technologies for robotic and manned missions; (2) the construction of a spaceport near the Earth and an infrastructure for the transport of interplanetary probes and staff for the manned exploration of the Solar System; and finally (3) there will be the actual exploration and colonization of nearby planets by humans.

Phase 1, dedicated to propulsion systems, comprises the application to space of a larger programme of research and development of energy sources on Earth, from programmes to develop nuclear fission and fusion (or a mixture of the two), to the study of the exploitation of antimatter. This work might last a decade and would have an estimated cost of a few billion dollars. The great unknown is how much further we will be able to reduce the cost of producing antimatter, which is currently incalculable.

Today, the laws of the market do not yet apply to antimatter as an energy source, in the same way that a century ago no one was interested in giving a market price of uranium. We can only say that as the capacity for production increases, the price falls. We can estimate 20–30 billion dollars for the cost of a programme aimed at the production of antimatter, to be included within the basic energy research programme.

Phase 2 will put the greatest demand on public funds. In that phase we will have to design and build the spaceport, which could be in Earth orbit or, perhaps better, in a libration point between the Earth and the Moon. Then, having devised a means of propulsion, we can send manned missions to explore Sphere +3, and begin the robotic exploration of Sphere +4, in order to finalize the development of the propulsion of a starship for the manned exploration of Sphere +4 (like mine *Per aspera ad astra*).

It is almost impossible to put numbers and times to Phase 2. The manned missions to Mars and beyond (to the asteroids or Europa) will certainly be more expensive commitments. It could be done in 30 years after the first phase, and the cost over 30 years would not be less than 1 trillion dollars. However, we always talk of sums that could easily be covered by a concomitant reduction in public spending on armaments by just a few percent.

And now we come to the third phase, which begins with the decision to leave for a nearby extra-solar planet in a high-speed, manned ship. The decision would be made under the pressure of public opinion once exploratory missions in Phase 2 had identified a really interesting planet. At this stage, the private sector comes into play; their contribution has the historical precedent of the colonization of North

America, which took place under a strategic and economic incentive. While London launched its privateers against the rich Spanish galleons carrying the treasures of the colonies, it also aimed to break the Spanish (and partly French) monopoly of the market in new, interesting goods that were coming from the Americas. Establishing English colonies seemed to be the ideal solution, but the passage across the Atlantic was especially difficult and expensive: the settlers, who were men and women willing, for various reasons, to migrate to the new world could not afford it. "Companies" were created (the first was the Virginia Company) with private capital to finance (and rule) the colonies. The most spectacular case was that of the Massachusetts colony founded by the Pilgrim Fathers—a group of Puritans who had broken away from the Anglican Church. The journey of the Pilgrims on the famous *Mayflower*, in 1620, was funded through an agreement with a leading London merchant, Thomas Weston, who gathered in considerable capital from London businessmen on the promise of attractive profits. It took 7 years (of torment, death, terrible winters, native Indians, etc.) before the few surviving colonists could repay their debts to the tough businessmen in London; but they did and established a principle that quickly became general during the populating of the future United States. The private financing of the American colonies is a classic example in the history of economics, which is especially instructive because it was applied when the public resources of the British Crown were stretched to the limit (and perhaps beyond it) because of the construction and maintenance of an extensive naval fleet, and the continuing wars on the Continent. Very little of the (high) English taxes, destined to support a military empire, could be used for a civilian enterprises. Individuals were the only solution, and they did it; although in 1620, they did not know how it would all turn out a century and a half later against those bellicose American colonists. They had a revolution, and by winning a bloody war of independence they quickly ousted from their new homeland the throne of their former rulers who were then the foremost military and industrial power in the world.

The colonization of another planet could follow the same pattern (although not necessarily with a final revolution; but you never know). But there would be one important advantage, for which the lenders will have to thank Einstein. For the bold explorers traveling at a significant fraction of the speed of light, a round trip will last, say, 30 years, while on Earth perhaps 70 years will pass. The scheme would be as follows: a group of persons sufficiently wealthy and willing to invest, even in such a risky enterprise (such as those who financed the *Mayflower*, which could very well have sunk), would found the Deep Space Company (DSC).

The DSC would agree to finance the expedition for, say, 1 zillion dollars (1 Z USD), or one quarter of the total cost (4 Z USD) of our hypothetical ship and its fuel. In return, they would get certain rights to the new planet, under an agreement with the "settlers" (which we assume to be brave but penniless). The settlers, for their part, in a rather pathetic attempt at *crowd financing* (a fancy way of saying that they go around with a begging bowl) procure the money for the equipment, etc.. At this point, the DSC enter into an agreement with one or more large banks, according to which they deposit another quarter (another 1 Z USD) of the value of the vessel for the duration of the trip (i.e., 70 Earth years), earning 2 % per annum on the deposit. In the end, the value of the deposit will be more than 70 Z USD, given the starting capital. The DSC, contemporaneously with the interest-bearing deposit, issues bonds tax at a zero rate, (otherwise known as zero-coupon bonds), lasting 70 years, guaranteed by the deposit.

A zero rate bond (or zero-coupon bond) is a bond bought at a lower price than its face value, while its nominal value is repaid in full at maturity. It gives no interest before maturity (zero-coupon). The DSC decides that the purchase price of the zero-coupon bond as issued will be 3 Z USD, but its face value at maturity (after 70 years) will be about 50 Z USD, which would be well-secured by the original deposit, plus interest (i.e., 70 Z USD).

Everyone is happy: the DSC which finances everything has a guaranteed profit; those who bought the zero-coupon bonds have a guaranteed investment which will be profitable in the long-term (to their heirs), and will certainly come with attractive tax benefits; and the scouts or settlers have the opportunity of their great adventure. Not to mention the fact that the contract states that, on its return, the DSC becomes the owner of the starship (which, after some maintenance can be made ready for the next trip) and, moreover, the DSC also expects profits from the colonies. And if the explorers have also invested during their time away, remember that for the calculation of interest on their capital, while 70 years have passed on the Earth, for them only 30 years will have passed … a further unexpected gain as "the Einstein Factor" makes its first appearance in the field of economics.

Printed in the United States
By Bookmasters